Tucholsky   Wagner   Zola   Scott   Schlegel
     Turgenev   Wallace   Fonatne Sydow   Freud
        Twain   Walther von der Vogelweide   Fouque   Friedrich II. von Preußen
                    Weber   Freiligrath
  Fechner   Weiße Rose   Kant   Ernst   Frey
          Fichte   von Fallersleben   Richthofen   Frommel
              Engels   Fielding   Hölderlin
       Fehrs   Faber   Flaubert   Eichendorff   Tacitus   Dumas
                  Maximilian I. von Habsburg   Eliasberg   Ebner Eschenbach
     Feuerbach   Fock   Eliot   Zweig
                         Ewald   Vergil
              Goethe   Elisabeth von Österreich   London
   Mendelssohn   Balzac   Shakespeare   Ganghofer
                      Lichtenberg   Rathenau   Dostojewski
          Trackl   Stevenson   Doyle   Gjellerup
  Mommsen   Tolstoi   Hambruch
          Thoma   Lenz   Hanrieder   Droste-Hülshoff
       Dach   Verne   von Arnim   Hägele   Hauff   Humboldt
              Reuter   Rousseau   Hauptmann   Gautier
          Karrillon   Garschin   Hagen   Baudelaire
                          Defoe   Hebbel
              Damaschke   Descartes   Hegel   Kussmaul   Herder
  Wolfram von Eschenbach   Dickens   Schopenhauer   Rilke   George
                     Darwin   Grimm Jerome   Bebel
       Bronner   Melville   Proust
                Campe   Horváth   Aristoteles   Federer
    Bismarck   Vigny   Barlach   Voltaire   Herodot
                              Gengenbach   Heine
        Storm   Casanova   Tersteegen   Grillparzer   Georgy
                     Chamberlain   Lessing   Langbein   Gilm   Gryphius
    Brentano   Lafontaine
       Strachwitz   Claudius   Schiller   Kralik   Iffland Sokrates
                                     Bellamy   Schilling
              Katharina II. von Rußland   Gerstäcker   Raabe   Gibbon   Tschechow
        Löns   Hoffmann   Gogol   Wilde   Vulpius
              Hesse   Gleim
         Luther   Heym   Hofmannsthal   Morgenstern
            Roth   Heyse   Klopstock   Klee   Hölty   Goedicke
   Luxemburg   Puschkin   Homer   Kleist
                              La Roche   Horaz   Mörike
              Machiavelli   Musil
     Navarra   Aurel   Musset   Kierkegaard   Kraft   Kraus
                                                                       Moltke
            Nestroy   Marie de France   Lamprecht   Kind   Kirchhoff   Hugo
                                Laotse   Ipsen   Liebknecht
              Nietzsche   Nansen   Ringelnatz
                          Marx   Lassalle   Gorki   Klett   Leibniz
         von Ossietzky   May   vom Stein   Lawrence
           Petalozzi   Knigge   Irving
                    Platon   Pückler   Kafka
              Sachs   Poe   Michelangelo   Kock
                                              Liebermann   Korolenko
                 de Sade   Praetorius   Mistral   Zetkin

The publishing house tradition has created the series **TREDITION CLASSICS**. It contains classical literature works from over two thousand years. Most of these titles have been out of print and off the bookstore shelves for decades.

The book series is intended to preserve the cultural legacy and to promote the timeless works of classical literature. As a reader of a **TREDITION CLASSICS** book, the reader supports the mission to save many of the amazing works of world literature from oblivion.

The symbol of **TREDITION CLASSICS** is Johannes Gutenberg (1400 – 1468), the inventor of movable type printing.

With the series, tradition intends to make thousands of international literature classics available in printed format again – worldwide.

All books are available at book retailers worldwide in paperback and in hardcover. For more information please visit: www.tredition.com

tradition was established in 2006 by Sandra Latusseck and Soenke Schulz. Based in Hamburg, Germany, tradition offers publishing solutions to authors and publishing houses, combined with worldwide distribution of printed and digital book content. tradition is uniquely positioned to enable authors and publishing houses to create books on their own terms and without conventional manufacturing risks.

For more information please visit: www.tredition.com

The Bounty of the Chesapeake
Fishing in Colonial Virginia

James Wharton

Imprint

This book is part of the TREDITION CLASSICS series.

Author: James Wharton
Cover design: toepferschumann, Berlin (Germany)

Publisher: tradition GmbH, Hamburg (Germany)
ISBN: 978-3-8491-4886-7

www.tredition.com
www.tredition.de

Copyright:
The content of this book is sourced from the public domain.

The intention of the TREDITION CLASSICS series is to make world literature in the public domain available in printed format. Literary enthusiasts and organizations worldwide have scanned and digitally edited the original texts. tredition has subsequently formatted and redesigned the content into a modern reading layout. Therefore, we cannot guarantee the exact reproduction of the original format of a particular historic edition. Please also note that no modifications have been made to the spelling, therefore it may differ from the orthography used today.

FOREWORD

Just as a series of personal letters may constitute an autobiography, so the extracts from Colonial writings that follow tell the unique story of the fisheries of Virginia's great Tidewater. In them it is possible to trace the measured growth of a vital industry. The interspersed comments of the compiler are to be understood as mere annotations. This is the testimony, then, of those who from the beginning participated in one of the foremost natural resources of this country.

I gratefully acknowledge guidance in research to Mr. John C. Pearson of the U.S. Fish & Wildlife Service, who masterfully surveyed the field and first brought the early fishery reports to public notice.

James Wharton

Weems, Virginia

THE BOUNTY OF THE CHESAPEAKE

The Bounty of The Chesapeake

The voyage to America in 1607 was like a journey to a star. Veteran rovers though the English were, none of them had any clear idea of what to expect in the new land of Virginia. Only one thing was certain: they would have nothing there but what they took with them or wrought from the raw materials of the country.

What raw materials?

They had reliable information that the climate was mild. Therefore, crops could be raised. They learned of inexhaustible timber: so ships and dwellings and industrial works could be built. They hoped for gold and dreamed of access to uncharted lands of adventure. But putting first things first, how would they eat in the meantime?

When Sir Walter Raleigh established the first English colony in "Virginia"—on what is now Roanoke island, North Carolina—two good reporters, one a writer, the other an illustrator, were commissioned to describe what they saw. This was twenty-two years before Jamestown and naturally all the material consisted of Indian life and customs. Thomas Hariot wrote:

For four months of the year, February, March, April and May, there are plenty of sturgeon; and also in the same months of herrings, some of the ordinary bigness as ours in England, but the most part far greater, of eighteen, twenty inches, and some two feet in length and better; both these kinds of fish in these months are most plentiful and in best season which we found to be most delicate and pleasant meat.

There are also trouts, porpoises, rays, oldwives, mullets, plaice, and very many other sorts of excellent good fish, which we have taken and eaten, whose names I know not but in the country language we have of twelve sorts more the pictures as they were drawn in the country with their names.

The inhabitants use to take them two manner of ways, the one is by a kind of weir made of reeds which in that country are very strong. The other way which is more strange, is with poles made sharp at one end, by shooting them into the fish after the manner as Irishmen cast darts; either as they are rowing in their boats or else as they are wading in the shallows for the purpose.

There are also in many places plenty of these kinds which follow:

Sea crabs, such as we have in England.

Oysters, some very great, and some small; some round and some of a long shape. They are found both in salt water and brackish, and those that we had out of salt water are far better than the other as in our own country.

Also mussels, scallops, periwinkles and crevises.

Seekanauk, a kind of crusty shellfish which is good meat about a foot in breadth, having a crusty tail, many legs like a crab, and her eyes in her back. They are found in shallows of salty waters; and sometimes on the shore.

There are many tortoises both of land and sea kind, their backs and bellies are shelled very thick; their head, feet and tail, which are in appearance, seem ugly as though they were members of a serpent or venomous; but notwithstanding they are very good meat, as also their eggs. Some have been found of a yard in breadth and better.

In a charming drawing of a group of Indian maidens John White, the artist associate, commented: "They delight ... in seeing fish taken in the rivers."

Over and over the first visitors to the Chesapeake bay painted rosy pictures of its marine life, stressing the abundance, variety and tastiness of the fish and shellfish. Exploration and communication were chiefly by water: it was natural that emphasis be laid on water resources. Though it is proverbial that fish stories partake of fiction, in the case of John Smith and his successors, it is doubtful whether they were greatly exaggerated. This was a world where nature, especially in the waters, was immeasurably prolific.

On the other hand, the conclusions drawn by many of those reading the reports were probably unjustified. The infinite plenty was one thing. Making constant and profitable use of it was another.

Thus, although Smith cited an impressive roster of edible fish in the vicinity of Jamestown, it was not to follow that the settlers were always able to turn them to advantage. There were several good reasons.

Long before Jamestown the fisheries off the coast of Northern America and Canada were known to be richly productive, with promise of an organized and dependable industry. But farther south conditions were found to be quite different. The fishing in the Chesapeake bay had frustrating ways. Sometimes there were hordes of fish. Again they stayed away in large numbers. They were usually present during warm weather when spoilage was worst. The first colonists had no ice at all and very little salt. Frequent spells of damp weather made sun-drying impractical. If more fish were caught than could be eaten at once, the excess was very likely wasted. Fishing gear was consistently inadequate. But from the very first, fishing and its development had been kept in mind by the promoters of the colony.

Fishing rights were defined in 1606 in letters patent to Sir Thomas Gates, Sir George Somers and others, as recorded in the Charter granted in 1606:

They shall have all ... fishings ... from the said first seat of their plantation and habitation by the space of fifty miles of English statute measure, all along the said coast of Virginia and America, towards the west and southwest, as the coast lies ... and also all ... fishings for the space of fifty English miles ... all along the said coast of Virginia and America, towards the east and northeast ... and also ... fishings ... from the same, fifty miles every way on the sea coast, directly into the mainland by the space of one hundred like English miles.

In the new fishing territory around Jamestown the Indians were the professionals and their methods were of great interest to the English novices. A description is furnished by William Strachey, secretary of state of the colony and author of *The Historie of Travaile into Virginia Britannia*:

Their fishing is much in boats. These they call quintans, as the West Indians call their canoas. They make them with one tree, by burning and scraping away the coals with stones and shells till they have made them in the form of a trough. Some of them are an ell deep and forty or fifty foot in length and some will transport forty men, but the most ordinary are smaller and will ferry ten or twenty, with some luggage, over their broadest rivers. Instead of oars, they use paddles and sticks, with which they will row faster than we in our barges. They have nets for fishing, for the quantity as formerly braided and meshed as ours and these are made of bark of certain trees, deer sinews, or a kind of grass, which they call pemmenaw, of which their women between their hands and thighs, spin a thread very even and readily, and this thread serves for many uses, as about their housing, their mantles of feathers and their [?] and they also with it make lines for angles.

Their angles are long small rods at the end whereof they have a cleft to which the line is fastened, and at the line they hang a hook, made either of a bone grated (as they nock their arrows) in the form of a crooked pin or fishhook, or of the splinter of a bone, and with a thread of the line they tie on the bait. They use also long arrows tied on a line, wherewith they shoot at fish in the rivers. Those of Accowmack use staves, like unto javelins, headed with bone; with these they dart fish, swimming in the water....

By their houses they have sometimes a scaena or high stage, raised like a scaffold, or small spelts, reeds, or dried osiers covered with mats which gives a shadow and is a shelter ... where on a loft of hurdles they lay forth their corn and fish to dry....

They are inconstant in everything but what fear constrain them to keep; crafty, timorous, quick of apprehension, ingenious enough in their own works, as may testify their weirs in which they take their fish, which are certain enclosures made of reeds and framed in the fashion of a labyrinth or maze set a fathom deep in the water with divers chambers or beds out of which the entangled fish cannot return or get out, being once in. Well may a great one by chance break the reeds and so escape, otherwise he remains a prey to the fishermen the next low water which they fish with a net at the end of a pole....

The earliest observers reveal how intimately food from the waters was linked with the colonists' experiences. George Percy wrote in 1607:

We came to a place [Cape Henry] where they [natives] had made a great fire and had been newly roasting oysters. When they perceived our coming, they fled away to the mountains and left many of the oysters in the fire. We ate some of the oysters which were very large and delicate in taste.

This was April 27 of that year. Oyster roasts have been a Virginia institution ever since. He continued:

Upon this plot of ground [Lynnhaven Bay] we got good store of mussels and oysters, which lay on the ground as thick as stones. We opened some and found in many of them pearls.

The pearls would probably not have been worth mentioning, except as a novelty, if they had come from oysters alone. The Virginia oyster pearl lacks luster. But the mussel, particularly the one found in the James river, yields an iridescent pearl of some little value.

A month later more oysters, in a form unknown in Virginia today, were obtained from Indians by Captain Christopher Newport in return for ornaments, according to Gabriel Archer in 1607:

He notwithstanding with two women and another fellow of his own consort followed us some six miles with baskets full of dried oysters and met us at a point, where calling to us, we went ashore and bartered with them for most of their victuals.

A letter from the Council in Virginia to the Council in England in 1607 stated:

We are set down eighty miles within a river, for breadth, sweetness of water, length navigable up into the country, deep and bold channel, so stored with sturgeon and other sweet fish as no man's fortune has ever possessed the like. And, as we think, if more may be wished in a river it will be found.

After various vicissitudes John Smith confessed:

Though there be fish in the sea, fowls in the air, and beasts in the woods, their bounds are so large, they so wild, and we so weak and ignorant, we cannot much trouble them.

George Percy introduced a happier note:

It pleased God, after a while, to send those people which were our mortal enemies [Indians] to relieve us with victuals, as bread, corn, fish, and flesh in great plenty, which was the setting up of our feeble men, otherwise we had all perished.

John Smith tells about another crisis:

Our victuals being within eighteen days spent and the Indians' trade decreasing, I was sent to the mouth of the river, to Kecoughtan [Hampton], an Indian town, to trade for corn and try the river for fish, but our fishing we could not effect by reason of the stormy weather.... Only of sturgeon we had great store, whereon our men would so greedily surfeit, as it cost many their lives.

And still another:

From May to September, those that escaped lived upon sturgeon and sea crabs.

And this:

So it happened that neither we nor they had anything to eat but what the country afforded naturally. Yet of eighty who lived upon oysters in June or July, with a pint of corn a week for a man lying under trees, and one hundred twenty for the most part living upon sturgeon, which are dried till we pounded it to powder for meal, yet in ten weeks but seven died.

For once he paints a brighter picture:

The next night, being lodged at Kecoughtan, six or seven days the extreme wind, rain, frost, and snow caused us to keep Christmas among the savages, where we were never more merry, nor fed on more plenty of good oysters, fish, flesh, wild fowl, and good bread.

He describes further ups and downs:

Now we so quietly followed our business that in three months, we ... provided nets and weirs for fishing.

Sixty or eighty with Ensign Laxon were sent down the river to live upon oysters, and twenty with Lieutenant Percy to try fishing at Point Comfort. But in six weeks, they would not agree once to cast out their net.

We had more sturgeon than could be devoured by dog or man, of which the industrious by drying and pounding, mingled with caviar, sorrel, and other wholesome herbs, would make bread and good meat.

Despite the privations much food is available, Smith avers:

In summer no place affords more plenty of sturgeon, nor in winter more abundance of fowl, especially in time of frost. There was once taken fifty-two sturgeon at a draught, at another draught sixty-eight. From the latter end of May till the end of June are taken few but young sturgeon of two foot or a yard long. From thence till the midst of September them of two or three yards long and a few others. And in four or five hours with one net were ordinarily taken seven or eight; often more, seldom less. In the small rivers all the year there is a good plenty of small fish, so that with hooks those that would take pains had sufficient....

Of fish we were best acquainted with sturgeon, grampus, porpoise, seals, stingrays whose tails are very dangerous, brits, mullets, white salmon, trouts, soles, plaice, herring, conyfish, rockfish, eels, lampreys, catfish, shad, perch of three sorts, crabs, shrimps, crevises, oysters, cockles, and mussels. But the most strange fish is a small one so like the picture of St. George's dragon as possibly can be, except his legs and wings; and the toadfish which will swell till it be like to burst when it comes into the air.

When Smith spoke of sturgeon he was most probably referring to the James river, the best waters for sturgeon in Virginia to this day. The "small rivers" were the fresh-water tributaries of the large salty ones. The small fish to be found there which would take the hook in winter were probably the non-migratory species like perch, catfish and suckers. If some of the names Smith gives seem puzzling today, it should be remembered that often the same fish name has applied throughout history to different fish at different times or in different areas. Contrariwise, different names, in regional usage, may apply to the same fish. Thus it is virtually impossible to say whether all the fish named by Colonial reporters are to be found in Virginia waters today. For example, though no "white salmon" are known in Virginia, it is possible that Smith referred to a fish that merely resembled a salmon without belonging to that family. On the other hand, it is conceivable that Virginia boats caught "white salmon" in the Atlantic Ocean. "Conyfish" can mean several different fishes, so that it is not possible to be sure what Smith had in mind; so with "brit." "Crevise" is an older name for crawfish. Seals still make rare appearances in the bay. As for the stingrays, he spoke from experience; he was spiked by one. Almost all of his list are still being caught off Jamestown. The "St. George's dragon" or sea horse, is among them.

There are many more varieties of fish caught by Virginia fishermen today than were ever mentioned in Colonial records. This is due to superior gear and the more intensive use of it.

Captain Christopher Newport was among the earliest observers confirming Smith. He wrote in 1607:

The main river [James] abounds with sturgeon, very large and excellent good, having also at the mouth of every brook and in every creek both store and exceedingly good fish of divers kinds. In the large sounds near the sea are multitudes of fish, banks of oysters, and many great crabs rather better, in fact, than ours and able to suffice four men. And within sight of land into the sea we expect at time of year to have a good fishing for cod, as both at our entering we might perceive by palpable conjectures, seeing the cod follow the ship ... as also out of my own experience not far off to the northward the fishing I found in my first voyage to Virginia....

The commodities of the country, what they are in else, is not much to be regarded, the inhabitants having no concern with any nation, no respect of profit.... Yet this for the present, by the consent of all our seamen, merely fishing for sturgeon cannot be worth less than £1,000 a year, leaving herring and cod as possibilities....

We have a good fishing for mussels which resemble mother-of-pearl, and if the pearl we have seen in the king's ears and about their necks come from these shells we know the banks.

The crab "able to suffice four men" could scarcely have been other than the horseshoe. It has never been considered a delicacy.

It is usually by contraries that the truth is determined. Even in the midst of the apparent plenty of fish, fishing crews sometimes came home empty-handed after continued effort. Often storms interfered.

From personal experience John Smith was able to sound the warning about Chesapeake weather:

Our mast and sail blew overboard and such mighty waves overraked us in that small barge that with great danger we kept her from sinking by freeing out the water.

The winds are variable, but the like thunder and lightning to purify the air I have seldom either seen or heard in Europe.

As if struck by the helplessness of the settlers, a compassionate chief extended aid to them in 1608. A letter from Francis Perkins tells the story:

So excessive are the frosts that one night the river froze over almost from bank to bank in front of our harbour, although it was there as wide as that of London. There died from the frost some fish in the river, which when taken out after the frost was over, were very good and so fat that they could be fried in their own fat without adding any butter or such thing....

Their own great emperor or the wuarravance, which is the name of their kings, has sent some of his people that they may teach us how to sow the grain of this country and to make certain traps with which they are going to fish.

A letter from the Council in Virginia to the Virginia Company in London in 1610 shows that such favors were returned:

Whilst we were fishing divers Indians came down from the woods unto us and ... I gave unto them such fish as we took ... for indeed at this time of the year [July] they live poor, their corn being but newly put into the ground and their own store spent. Oysters and crabs and such fish as they take in their weirs is their best relief.

Oysters occurred in vast banks and shoals within sight of the Jamestown fort. During the 1609-10 "starving time" a minimum force was retained at the settlement while everyone else was turned out to forage as best he could. Most sought the oyster grounds where they ate oysters nine weeks, a diet varied only by a pitifully negligible allowance of corn meal. In the words of one of the foragers, "this kind of feeding caused all our skin to peel off from head to foot as if we had been dead." The arrival of supplies ended the ordeal. But soon hunger descended again and the oyster beds would have been the natural recourse if it had not been winter and the water too cold to wade in. So the oysters were no help.

That conscientious reporter, William Strachey, wrote in 1610:

In this desolation and misery our Governor found the condition and state of the Colony. Nor was there at the fort, as they whom we found related unto us, any means to take fish; neither sufficient seine, nor other convenient net, and yet of their need, there was not one eye of sturgeon yet come into the river.

The river which was wont before this time of the year to be plentiful of sturgeon had not now a fish to be seen in it, and albeit we laboured and hauled our net twenty times day and night, yet we took not so much as would content half the fishermen. Our Governor therefore, sent away his long boat to coast the river downward as far as Point Comfort, and from thence to Cape Henry and Cape Charles, and all within the bay, which after a seven nights trial and travail, returned without any fruits of their labours, scarce getting so much fish as served their own company.

And, likewise, because at the Lord Governor and Captain General's first coming, there was found in our own river no store of fish after many trials, the Lord Governor and Captain General dispatched in the *Virginia*, with instructions, the seventeenth of June, 1610, Robert Tyndall, master of the *De la Warre*, to fish unto, all along, and between Cape Henry and Cape Charles within the bay.... Nor was the Lord Governor and Captain General in the meanwhile idle at the fort, but every day and night he caused the nets to be hauled, sometimes a dozen times one after another. But it pleased not God so to bless our labours that we did at any time take one quarter so much as would give unto our people one pound at a meal apiece, by which we might have better husbanded our peas and oatmeal, notwithstanding the great store we now saw daily in our river. But let the blame of this lie where it is, both upon our nets and the unskilfulness of our men to lay them.

The matter of sturgeon was of prime importance not only for subsistence but for export, particularly of the roe. Caviar was in great demand in England. But with uncertainty as to when the sturgeon would appear in the river, plus hot weather, plus feeble facilities, the growth of the industry was impeded. When tobacco, first commercially grown by John Rolfe, appeared on the scene in 1612 and proved to be a sure money maker, the export of sturgeon products came to a standstill. It was having hard going anyway. Complaints from England regarding quality were familiar enough. According to Lord De La Warr in 1610, on the subject, "Virginia Commodities":

> Sturgeon which was last sent came ill-conditioned, not being well boiled. If it were cut in small pieces and powdered, put up in cask, the heads pickled by themselves, and sent here, it would do far better.

> Roes of the said sturgeon make caviar according to instructions formerly given. Sounds of the said sturgeon will make isinglass according to the same instructions. Isinglass is worth here 13s. 4d. per 100 pounds, and caviar well conditioned is worth £40 per 100.

Other instances stressed the undependable fishing. Lord De La Warr wrote to the Earl of Salisbury in England in 1610: "I sent fishermen out to provide fish for our men, to save other provision, but they had ill success."

Captain Samuel Argall was specially commissioned by the authorities in England to deep-sea fish for the benefit of the Colony. After ranging over a wide area between Bermuda and Canada, he reported in 1610:

> ... The weather continuing very foggy, thick, and rainy, about five of the clock it began to cease and then we began to fish and so continued until seven of the clock in between thirty and forty fathoms, and then we could fish no longer. So having gotten between twenty and thirty cods we left for that night, and at five of the clock, the 26th, in the morning we began to fish again and so continued until ten of the clock, and then it would fish no longer, in which time we had taken near one hundred cods and a couple of halibuts....

> Then I tried whether there were any fish there or not [off Maine coast], and I found reasonable good store there. So I stayed there fishing till the 12th of August, [1610] and then finding that the fishing did fail, I thought good to return to the island [Jamestown]....

Captain Argall also offered his opinion of the usefulness of the islands off Virginia's seacoast peninsula, later known as the Eastern Shore:

Salt might easily be made there, if there were any ponds digged, for that I found salt kernel where the water had overflowed in certain places. Here also is great store of fish, both shellfish and others.

The root of the trouble, so far as local fishing conditions were concerned, was the lack of adequate equipment together with ignorance of its proper use. Perhaps the ease with which fish were caught at certain times had spoiled the hardy settlers.

A low opinion of their attitude in this vital pursuit came from Sir Thomas Gates in 1610:

A colony is therefore denominated because they should be coloni, the tillers of the earth and stewards of fertility. Our mutinous loiterers would not sow with providence and therefore they reaped the fruits of far too dear bought repentance. An incredible example of their idleness is the report of Sir Thomas Gates who affirms that after his first coming thither he had seen some of them eat their fish raw rather than they would go a stone's cast to fetch wood and dress it.

Joined unto these another evil: There is great store of fish in the river, especially of sturgeon, but our men provided no more of them than present necessity, not barreling up any store against the season [when] the sturgeon returned to the sea. And not to dissemble their folly, they suffered fourteen nets, which was all they had, to rot and spoil, which by orderly drying and mending might have been preserved but being lost, all help of fishing perished.

Very few of them had come equipped for fishing. Their seines were as old-fashioned as those used by the Apostles in the New Testament, the simple kind you lowered from a boat and dragged ashore. The Indians had taught them how to spear large fish and erect weirs out of stakes and brushwood to entrap migrating schools. Such methods worked well enough during the season. But in cold weather, when provisions ran low, scarcely any fish were present in the bay proper.

It was different in New England and Canada. There the fishing was good the year round. The sea bottom was dragged by efficient trawl-nets, and fished with gang-lines of baited hooks, as it still is today. The cool temperatures over many months of the year made the catches much less perishable. Conditions favored an organized fish-salting industry.

Though the Jamestown people had easy access to some 3,000 square miles of inland tidal water and were only a little way from the open sea, they never developed their marine riches. One good reason was that their original aims were in other directions. When the first intentions to colonize New England came to the King's notice, he asked the leaders what drew them there. The one-word answer: "Fishing." If the Virginians had been similarly queried they would have given various replies, but certainly not that one.

In describing the fisheries of New England, John Smith had enthused:

Let not the meanness of the word fish distaste you, for it will afford us good gold as the mines of Guiana or Tumbata, with less hazard and charge, and more certainty and facility.

The need for fishermen in Virginia was officially recognized to only a slight degree. A 1610 memorandum from the Virginia Council to the authorities in London asked that an effort be made to include among the next immigrants 20 fishermen and 6 net makers. Select them with care was the word sent out in England by means of a broadside issued by the Council of Virginia, December, 1610:

Whereas the good ship called the *Hercules* is now preparing and almost in a readiness with necessary provisions to make a supply to the Lord Governor and the Colony in Virginia, it is thought meet, for the avoiding of such vagrant and unnecessary persons as do commonly proffer themselves being altogether unserviceable, that none but honest sufficient artificers, as carpenters, smiths, coopers, fishermen, brickmen, and such like, shall be entertained into this voyage. Of whom so many as will in due time repair to the house of Sir Thomas Smith in Philpot Lane, with sufficient testimony to their skill and good behavior, they shall receive entertainment accordingly.

It was only a question of time before the Virginia colonists would, though surrounded all the while by their own huge marine resources, subsist on salt fish from the North. Sir Thomas Dale, governor from 1611 to 1616, perceived the trend. One of his first moves was to ask the President of the Virginia Company to provide men trained enough to build a coastal trade in furs, corn and fish:

Let me intreat that we may have both an admiral and hired mariners, to be all times resident here. The benefit will quickly make good the charge as well by a trade of furs to be obtained with the savages in the northern rivers to be returned home as also to furnish us here with corn and fish. The waste of such men all this time whom we might trust with our pinnaces leaves us destitute this season of so great a quantity of fish as not far from our own bay would sufficiently satisfy the whole Colony for a whole year.

There were no boats available even for simple oystering. During the term of the stringent Governor Dale some disaffected colonists tried to escape in a shallop and a barge, which were "all the boats that were then in the Colony."

Ironically punctuating the sagas of hardship were the marveling descriptions publicized in England. Corroborating the mouth-watering tales of Smith, William Strachey wrote in 1612:

To the natural commodities which the country has of fruit, beasts, and fowl, we may also add the no mean commodity of fish, of which, in March and April, are great shoals of herrings, sturgeon, great store commonly in May if the year be forward. I have been at the taking of some before Algernoone fort and in Southampton river in the middle of March, and they remain with us June, July, and August and in that plenty as before expressed.

Shad, great store, of a yard long and for sweetness and fatness a reasonable food fish; he is only full of small bones, like our barbels in England. There is the garfish, some of which are a yard long, small and round like an eel and as big as a mare's leg, having a long snout full of sharp teeth.

Oysters there be in whole banks and beds, and those of the best. I have seen some thirteen inches long. The savages use to boil oysters and mussels together and with the broth they make a good spoon meat, thickened with the flour of their wheat and it is a great thrift and husbandry with them to hang the oysters upon strings ... and dried in the smoke, thereby to preserve them all the year.

There be two sorts of sea crabs. One our people call a king crab and they are taken in shoal waters from off the shore a dozen at a time hanging one upon another's tail; they are of a foot in length and half a foot in breadth, having legs and a long tail. The Indians seldom eat of this kind. There is a shellfish of the proportion of a cockle but far greater [conch]. It has a smooth shell, not ragged as our cockles; 'tis good meat though somewhat tough.

And, according to Alexander Whitaker in 1613:

The rivers abound with fish both small and great. The sea-fish come into our rivers in March and continue the end of September. Great schools of herrings come in first; shads of a great bigness and the rockfish follow them. Trout, bass, flounders, and other dainty fish come in before the others be gone. Then come multitudes of great sturgeons, whereof we catch many and should do more, but that we want good nets answerable to the breadth and depth of our rivers. Besides our channels are so foul in the bottom with great logs and trees that we often break our nets upon them. I cannot reckon nor give proper names to the divers kinds of fresh fish in our rivers. I have caught with mine angle, carp, pike, eel, perches of six several kinds, crayfish and the torope or little turtle, besides many small kinds.

When Whitaker penned the word "torope," he was giving the English-speaking world a new term, new because the animal it defined was unknown in Europe. Later spelled "terrapin," it meant the diamond-back, the esoteric little creature that spread the fame of the Chesapeake bay around the world and became an indispensable course on menus designed for the entertainment of royalty and the discriminating elect. The colonists probably ate it prepared Indian fashion, that is, roasted whole in live coals and opened at table where the savory meat was extracted by appreciative fingers. Over generations of terrapin-fanciers it evolved into one of the stars of the gastronomic firmament. It is a wholly American dish and it was born at Jamestown.

Contemporary Historian Ralph Hamor added his testimony in 1614:

For fish, the rivers are plentifully stored with sturgeon, porpoise, bass, rockfish, carp, shad, herring, eel, catfish, perch, flat-fish, trout, sheepshead, drummers, jewfish, crevises, crabs, oysters, and divers other kinds. Of all which myself has seen great quantity taken, especially the last summer at Smith's Island at one haul a frigate's lading of sturgeon, bass, and other great fish in Captain Argall's seine, and even at the very place which is not above fifteen miles from Point Comfort. If we had been furnished with salt to have saved it, we might have taken as much fish as would have served us that whole year.

The mention of carp will interest those who believe carp to have been introduced into Virginia much later. The jewfish is common in more southern waters but there may well have been some strays in the Chesapeake. Although croakers, one of the bay's most abundant fish in modern times, are not mentioned, it would not be unreasonable to assume that they were included under "drummers." So with spot, a member of the drum family bearing a superficial resemblance to a bass or perch. The term "spot," as applied to a Virginia fish does not seem to have become current till the late 19th century.

An event of special interest to statisticians occurred in 1612. The first attempt made in the New World to require certain fish catches to be reported was among the regulations propounded by Governor Thomas Dale. The penalty for violation would shock today's delinquent record keepers:

All fishermen, dressers of sturgeon, or such like appointed to fish or to cure the said sturgeon for the use of the Colony, shall give a just and true account of all such fish as they shall take by day or night, of whatsoever kind, the same to bring unto the Governor. As also all such kegs of sturgeon or caviar as they shall prepare and cure upon peril for the first time offending herein of losing his ears, and for the second time to be condemned a year to the galleys, and for the third time offending to be condemned to the galleys for three years.

The years of trial and error fishing had brought their return in increased knowledge, according to John Rolfe in 1616:

About two years since, Sir Thomas Dale ... found out two seasons in the year to catch fish, namely, the spring and the fall. He himself took no small pains in the trial and at one haul with a seine caught five thousand three hundred of them, as big as cod. The least of the residue or kind of salmon trout, two foot long, yet he durst not adventure on the main school for breaking his net. Likewise, two men with axes and such like weapons have taken and killed near the shore and brought home forty [fish] as great as cod in two or three hours space....

There was a hint that the Virginia Company was interfering with free ocean fishing by claiming all the land to Newfoundland,—not that it was getting much out of it. One complaint as published in London sometime before February 22, 1615, in the anonymous tract, *The Trades Increase*, read:

The Virginia Company pretend almost all that main twixt it and Newfoundland to be their fee-simple, whereby many honest and able minds, disposed to adventure, are hindered and stopped from repairing to those places that they either know or would discover, even for fishing.

As a matter of fact, there was continuous wrangling in London over the fishing rights off the entire coast administered by the Virginia Company. The proposed settlers of the Northern Colony in New England had fishing uppermost in their minds and would have been glad to exclude fishermen coming from the Southern Colony. Minutes of meetings of the Company reveal how earnest was the struggle:

December 1, 1619. The last great general court being read, Mr. Treasurer acquainted them that Mr. John Delbridge, purposing to settle a particular colony in Virginia, desired of the Company that for defraying some part of his charge he might be admitted to fish at Cape Cod. Which request was opposed by Sir Ferdinando Gorges, alleging that he always favored Mr. Delbridge but in this he thought himself something touched that he should sue to this Company and not rather to him as the matter properly belonged to the Northern Colony to give liberty for fishing in that place, it lying within their latitude. This was answered by Mr. Treasurer that the Companies of the South and North Plantations are free of one another and that the patent is clear that each may fish within the territory of the other, the sea being free for both. If the Northern Company abridged them of this, they would take away their means and encouragement for sending out men. To which Sir Ferdinando Gorges replied that if he was not mistaken both the Companies were limited by the patents unto which he would submit. For the deciding whereof it is referred to the Council, who are of both Companies, to examine the patents tomorrow afternoon at the Lord Southampton's and accordingly to determine the dispute.

Two weeks later the Council gave its decision: Either Colony could fish within the bounds of the other. But this was by no means an end to the matter. The Northern Colony requested a new patent to resolve the disputes. With suggestions and counter-suggestions, the debate dragged on through the spring, summer and fall. About the time the Northern Colony had arranged to exclude the Southern Colony from free fishing, the King stepped in, declaring that "if anything were passed in the New England patent that might be prejudicial to the Southern Colony it was done without his knowledge and that he has been abused thereby by those that pretended otherwise to him." Finally, after a year-and-a-half of cross-purposes, agreement was reached:

June 18, 1621. There was a petition exhibited unto His Majesty in the name of the patentees and adventurers in the plantation of New England concerning some difference between the Southern and Northern Colonies, the said petition was by His Majesty referred to the consideration of the Lords. Their Lordships, upon the hearing and debating of the matter at large and by the consent of both Colonies, did establish and confirm two former orders, the one bearing date of the 16th of March 1620, agreed upon by the Duke of Lenox and the Earl of Arundell; the other of the 21st of July 1620 ordered by the Board whereby it was thought fit that the said colonies should fish at sea within the limits and bounds of each other reciprocally, with this limitation that it be only for the sustentation of the people of the Colonies there and for the transportation of people into either Colony. Further it was ordered at this time by their Lordships that they should have freedom of the shore for drying of their nets and taking and saving of their fish and to have wood for their necessary uses, by the assignment of the Governors at reasonable rates. Lastly the patent of the Northern Colony shall be renewed according to the premises, and those of the Southern plantation to have a sight thereof before it be engrossed and the former patent to be delivered into the hand of the patentees.

In an effort to encourage Virginians to salt their own fish, an order from London recommended the reopening of the old sea-water-evaporators on Smith's island, off Cape Charles, where salt had been produced in the first days. The Virginia Company advised the Governor and Council in 1620:

The last commodity, but not of least importance for health, is SALT: the works whereof having been lately suffered to decay; we now intending to restore in so great plenty, as not only to serve the Colony for the present, but as is hoped, in short time, the great fishings on those coasts, a matter of inestimable advancement to the Colony, do upon mature deliberation ordain as followeth: First, that you the Governor and Council, do chose out of the tenants for the Company, 20 fit persons to be employed in salt works, which are to be renewed in Smith's Island, where they were before; as also in taking of fish there, for the use of the Colony, as in former times was also done. These 20 shall be furnished out at the first, at the charges of the Company, with all implements and instruments necessary for those works. They shall have also assigned to each of them for their occupation or use, 50 acres of land within the island, to be land of the Company. The one moiety of salt, fish, and profits of the land shall be for the tenants, the other for us the Company, to be delivered into our store: and this contract shall be continued for five years.

The reply of Secretary of the Colony, John Pory, was something less than complacent:

The last commodity spoken of in your charter is salt; the works whereof, we do much marvel, you would have restored to their former use; whereas I will undertake in one day to make as much salt by the heat of the sun, after the manner used in France, Spain, and Italy, as can be made in a year by that toilsome and erroneous way of boiling sea water into salt in kettles as our people at Smith's Island hitherto accustomed. And therefore when you enter into this work, you must send men skillful in salt ponds, such as you may easily procure from Rochell, and if you can have none there, yet some will be found in Lymington, and in many other places in England. And this indeed in a short time might prove a real work of great sustenance to the Colony at home, as of gain abroad, here being such schools of excellent fish, as ought rather to be admired of such as have not seen the same, than credited. Whereas the Company do give their tenants fifty acres upon Smith's Island some there are that smile at it here, saying there is no ground in all the whole island worth the manuring.

Following this exchange, attempts at salt making, especially on the Eastern Shore where the waters were saltiest, were renewed. John Rolfe reported in 1621:

At Dale's Gift, being upon the sea near unto Cape Charles, about thirty miles from Kecoughtan, are seventeen inhabitants under command of Lieutenant Cradock. All these are fed and maintained by the Colony. Their labor is to make salt and catch fish....

Secretary Pory soon expressed his disagreement with the project in more than words and succeeded in effecting the removal of the salt works to a more convenient location. That this hardly fulfilled expectations is evidenced by a letter written in 1628 to the King by the Governor and Council:

Great likeliness of the certainty of bay salt, the benefit that will thereby accrue to the Colony will be great, and they shall willingly assist Mr. Capps in making his experiment, which, brought to perfection, will draw a certain trade to them. And they hope that the fishing upon their coasts will be very near as good as Canada.

Mr. Capps, a citizen of Accomack, had proposed that if the Colony would subsidize him he would undertake to supply it with salt from evaporated sea water. His offer was accepted and the enterprise set up. After waiting patiently and seeing little salt the Council took him to task. His plea was the familiar one of most operations that fail: lack of capital. He had worked hard, he said; he had all the firewood he needed, workmen were available, and the sun shone bright. The bottle-neck was too few evaporating pans. But apparently he had not won the Council's confidence. The Capps salt company was dissolved.

Another one sprang up about 30 years later under the sponsorship of Colonel Edmund Scarborough of Northampton County. Such was the public interest aroused by this influential man, who, among other distinctions, had been a Burgess between 1642 and 1659, that the importation of salt into the county was prohibited to encourage him. Finally, in 1666, this project was abandoned for reasons that remain obscure. Most probably the quality of the product was inferior.

The salt shortage continued despite other random attempts to alleviate it. For example, in 1660 one Daniel Dawen of Accomack was exempted from taxes and granted public funds for his "experiments of salt."

The trouble that attended obtaining salt in needed quantity and of satisfactory quality accompanied the development of Virginia right up to George Washington's time.

Despite all attempts to the contrary, reliance on salt fish from the North kept gaining. The General Assembly that had met in 1619 censured a Captain Warde for establishing a plantation in Virginia without asking anybody's permission. But when it was brought out that he had conveyed quantities of salt fish to the Colony from Canada on his ship he was forgiven. This captain was an important link between the Colony and the North. John Rolfe wrote to Sir Edwin Sandys in 1619:

Captain Warde in his ship went to Monhegan [island, Maine] in the Northern Colony in May and returned the latter end of July with fish which he caught there. He brought but a small quantity by reason he had but little salt. There were some Plymouth ships where he harbored, who made great store of fish which is far larger than Newland [Newfoundland] fish.

The Maine waters were far busier than those of Virginia. For more than a century vessels from half-a-dozen European nations had thronged there, even to Greenland, attracted by the fishing, and the furs available on the mainland. When some of the early experiments at colonization failed, fishing became all the more emphasized. There was usually excellent demand for the catches whether landed in Plymouth (England) or Plymouth (Massachusetts), Portugal, Holland, the West Indies or Virginia. These bold adventurers made use of the land in the New World only for drying, salting and barreling their fish. If conditions permitted, they transported them fresh, in a cargo commonly known as "corfish." Oil made from whale and cod was a profitable commodity.

Fishermen were the pioneers and explorers of America's first days just as the miners, trappers and traders were those of a later period.

The importance of fish was thus underlined. In addition, conceding the value to the untrained whites of Indians as fishermen, the 1619 Assembly agreed to a proposal that Indians to the limit of six be permitted to live in white settlements if they engaged in fishing for the benefit of the settlement. Indian methods were first described by Hariot of the Roanoke island colony:

They have likewise a notable way to catch fish in their rivers, for whereas they lack both iron and steel, they fasten unto their reeds, or long rods, the hollow tail of a certain fish like to a sea crab instead of a point, wherewith by night or day they strike fishes, and take them up into their boats. They also know how to use the prickles, and pricks of other fishes. They also make weirs, with setting up reeds or twigs in the water, which they so plant one with another, that they grow still narrower, and narrower. There was never seen among us so cunning a way to take fish withal, whereof sundry sorts as they found in their rivers unlike ours, which are also of a very good taste. Doubtless it is a pleasant sight to see the people, sometimes wading, and going sometimes sailing in those rivers, which are shallow and not deep, free from all care of heaping up riches for their posterity, content with their state, and living friendly together of those things which God of His bounty hath given unto them, yet without giving Him any thanks according to His deserts.

The most vivid and comprehensive description of Indian fishing was given by historian Robert Beverley. Though his work was not published until 1705, he dealt with an earlier period:

Before the arrival of the English there, the Indians had fish in such vast plenty that the boys and girls would take a pointed stick and strike the lesser sort as they swam upon the flats. The larger fish that kept in deeper water, they were put to a little more difficulty to take. But for these they made weirs, that is, a hedge of small rived sticks or reeds of the thickness of a man's finger. These they wove together in a row with straps of green oak or other tough wood, so close that the small fish could not pass through. Upon high water mark they pitched one end of this hedge and the other they extended into the river to the depth of eight or ten foot, fastening it with stakes, making cods out from the hedge on one side, almost at the end, and leaving a gap for the fish to go into them. These were contrived so that the fish could easily find their passage into those cods when they were at the gap, but not see their way out again when they were in. Thus if they offered to pass through, they were taken.

Sometimes they made such a hedge as this quite across a creek at high water and at low would go into the run, so contracted into a narrow stream, and take out what fish they pleased.

At the falls of the rivers where the water is shallow and the current strong, the Indians use another kind of weir thus made. They make a dam of loose stone, whereof there is plenty at hand, quite across the river, leaving one, two, or more spaces or trunnels for the water to pass through. At the mouth they set a pot of reeds, wove in form of a cone, whose base is about three foot [wide] and ten [foot] perpendicular, into which the swiftness of the current carries the fish and wedges them so fast that they cannot possibly return.

The Indian way of catching sturgeon, when they came into the narrow part of the rivers, was by a man's clapping a noose over their tails and by keeping fast his hold. Thus a fish, finding itself entangled, would flounce and often pull him under water. Then that man was counted a cockarouse, or brave fellow, that would not let go till with swimming, wading and diving, he had tired the sturgeon and brought it ashore. These sturgeon would also leap into their canoes in crossing the river, as many of them do still every year into the boats of the English.

They have also another way of fishing like those on the Euxine Sea, by the help of a blazing fire by night. They make a hearth in the middle of their canoe, raising it within two inches of the edge. Upon this they lay their burning lightwood, split into small shivers, each splinter whereof will blaze and burn end for end like a candle. 'Tis one man's work to tend this fire and keep it flaming. At each end of the canoe stands an Indian with a gig or point spear, setting the canoe forward with the butt end of the spear as gently as he can, by that means stealing upon the fish without any noise or disturbing of the water. Then they with great dexterity dart these spears into the fish and so take them. Now there is a double convenience in the blaze of this fire, for it not only dazzles the eyes of the fish, which will lie still glaring upon it, but likewise discovers the bottom of the river clearly to the fisherman, which the daylight does not.

Under Governor George Yeardley in 1616, there were 400 people at Jamestown and one old frigate, one old shallop and one boat belonging to the community. There were two boats privately owned. The boats best suited to local fishing, and the most easily available, were the Indian dugout canoes. Such was the size of the trees that it was possible to make them comparatively roomy, as Strachey noted.

Every passing year brought home to the steadily growing Colony the need of improving its fishing practices. Most nets had to be bought in England. Here is a London item from a 1623 *List of Subscribers and Subscriptions for Relief of the Colony*: "Richard Tatem will adventure [speculate] in cheese and fishing nets the sum of £30 sterling."

Jamestown had by 1624 begun to spawn little Jamestowns throughout the countryside. A census was ordered of all settlements. In January, 1625, there were 1209 white persons, and 23 negroes. This first American census listed, among general provisions, the stocks of salt fish. On hand at thirteen settlements was 58,380 pounds. James City had the largest supply, 24,880 pounds. Elizabeth City was next with 10,550 pounds. A community listed only as "Neck of Land" adjacent to Jamestown, consisting of perhaps ten dwellings and plantations, had 4,050 pounds. The smallest store, 450 pounds, was credited to another "Neck of Land" in Charles City. From the accumulated evidences of disorganized home fishing, coupled with the deficiency of salt, it is to be concluded that most of this supply had come from the Northern fishing grounds.

There were 40 boats of various sizes and uses listed in this census. For example, at Jamestown a "barque of 40 tons, a shallop of 4 tons and one skiff" were among the ten there.

A token of the stress resulting from inadequate fisheries even after 16 years of active colonization is this letter preserved in the records of the Virginia Company. A Virginia citizen named Arundle in 1623 wrote to his friend, Mr. Caning, in London:

The most evident hope from altogether starving is oysters, and for the easier getting of them I have agreed for a canoe which will cost me 6 livres sterling.

Emigrants had been advised not to leave for Virginia without some fishing equipment. In his *Travels*, John Smith had included the warning: "A particular of such necessaries as either private families or single persons shall have cause to provide to go to Virginia ... nets, hooks and lines must be added."

Records of the Virginia Company in London throw light on the extensiveness of the fish trade. Robert Bennett wrote from Virginia to Edward Bennett in London in 1623:

My last letter I wrote you was in the *Adam* from Newfoundland, which I hope you shall receive before this. God send her back in safety and this from Canada. I hope the fish will come to a good reckoning for victuals is very scarce in the country. Your Newfoundland fish is worth 30s. per hundred, your dry Canada [fish] £3, 10s. and the wet £5, 10s. per hundred. I do not know nor hear of any that is coming hither with fish but only the *Tiger* which went in company with the *Adam* from this place and I know the country will carry away all this forthwith.

And again from the records of the Company, this extract from *An Account of Sums Subscribed and Supplies Sent Since April*, dated July 23, 1623:

... We have received advice that from Canada there departed this last month a ship called Furtherance with above forty thousand of that fish which is little inferior to ling for the supply of the Colony in Virginia and that fish is worth not less than £600.

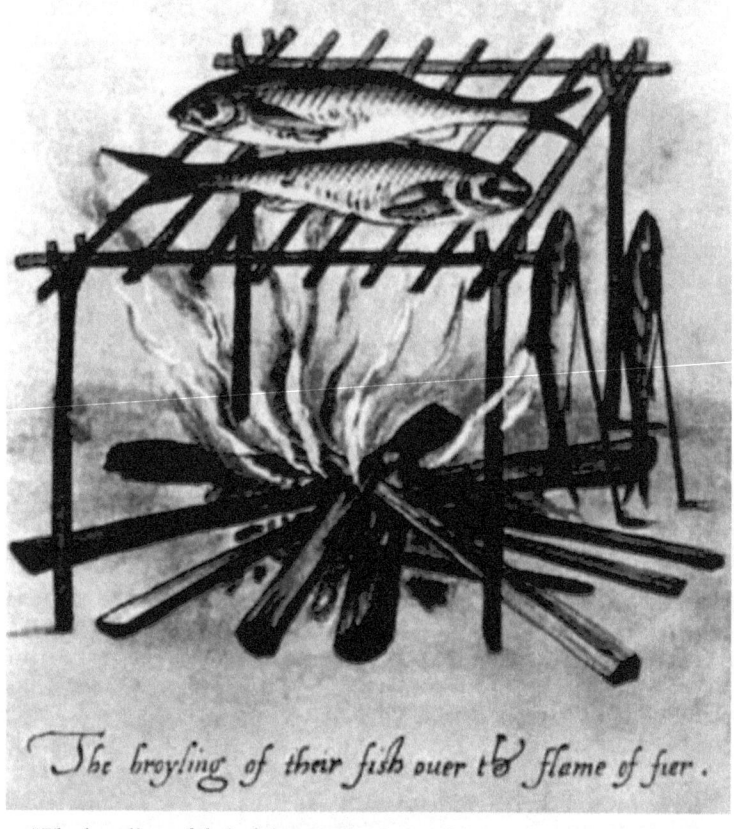

"The broyling of their fish over the flame of fire."

Library of Congress Photo

The first settlers did not have to learn from the Indians how to cook fish, but this method was perhaps as appetizing as any they knew.

The manner of their fishing.

Library of Congress Photo

The first colonists saw the Indians engaged in fishing practices that included spearing, luring with firelight, and entrapping in staked-off enclosures.

The sheepshead was one of the favorite seafoods of Tidewater Virginians from the beginning. It was fairly abundant, according to their records, and remained so until the twentieth century, when it became almost extinct in Chesapeake waters.

U.S. Fish and Wildlife Service Photos

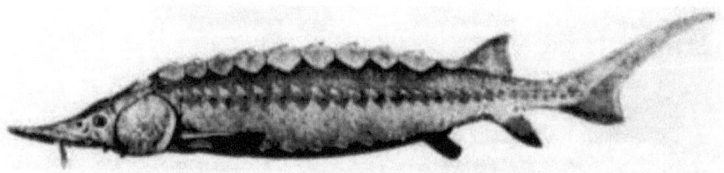

The ugly-looking but delicious-tasting sturgeon was the fish that principally engaged the attention of the first colonists. They were impressed by its abundance and were busy for a time in shipping its roe to England for [1]caviar.

U.S. Fish and Wildlife Service Photos

[1] (we cannot be certain that much actual caviar was produced at Jamestown. The chances are that the roe was merely salted down and that the final processing took place in England)

Haul-seining or dragging fish ashore by enclosing them in a long net, is a form of fishing that has thrived almost unchanged through the ages. Its practice at Jamestown was limited by the lack of nets.

The toothsome Chesapeake Bay hard crab was, and is still to a great extent today, taken by baits spaced along lines sunk to the bottom and then raised and the tenacious crabs removed.

Vast quantities of river herring were taken in haul-seines in the spring throughout Tidewater Virginia. A crew dragged the fish ashore to a force of women cutters waiting to prepare them for salting down.

Great living oyster mounds, built up by nature through the ages, impeded ships in the lower James river. At high tide they were hidden so that unwary pilots struck them; at low they could be picked over by hand. They remained a threat to navigation until they disappeared under three centuries of harvesting.

Original drawing by Esther Derieux

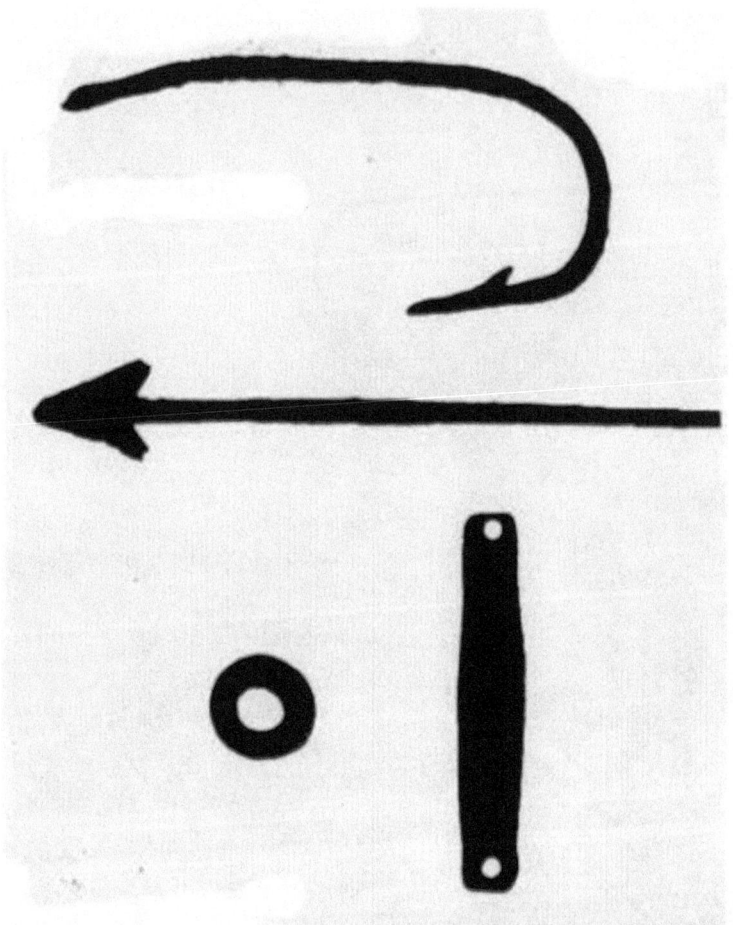

Fishing implements excavated at Jamestown. The large fish-hook was for ocean cod fishing or possibly for snagging sturgeon in the river. The spear, attached to a wooden handle, was for stalking big fish in shallow water, or for capturing those that could be attracted to a light in a boat at night. The lead weights were suitable for (right) a handline, (left) a net.

National Park Service

Early salt-evaporating houses were located close by the sea, from which the water was channeled in by slow stages to take advantage of natural evaporation before wood fires finished the job. When the crystals formed they were shoveled into conical baskets and drained.

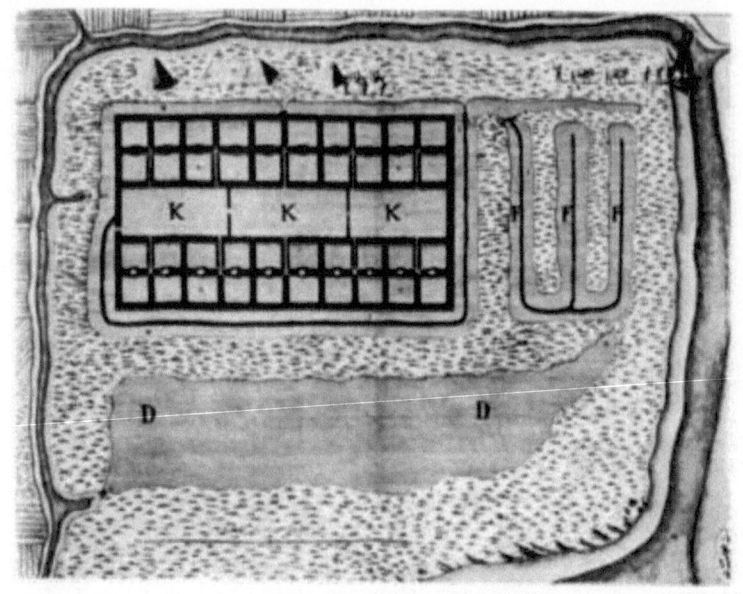

Courtesy Mariners Museum

An 18th century plan of a solar-evaporating works. Sea water is channeled into the primary reservoir (DD), from which it is conducted to (FFF) and (KKK) by progressive stages to the final basins where it crystallizes.

The kernel of the situation was reflected by the Dutch traveler, David De Vries, who made voyages to America from 1632 to 1644:

In going down to Jamestown on board of a sloop, a sturgeon sprang out of the river, into the sloop. We killed it, and it was eight feet long. This river is full of sturgeon, as also are the two rivers of New Netherland. When the English first began to plant their Colony here, there came an English ship from England for the purpose of fishing for sturgeon; but they found that this fishery would not answer, because it is so hot in summer, which is the best time for fishing, that the salt or pickle would not keep them as in Muscovy whence the English obtain many sturgeon and where the climate is colder than in the Virginias.

The effects of the Virginians' favoring tobacco-growing above fishing were also noted by De Vries on a visit to Canada:

Besides my vessel [at Newfoundland] there was a small boat of fifty or sixty lasts [110 tons], with six guns, which had come out of the Virginias with tobacco, in order to exchange the tobacco for fish.

A rather aggrieved reaction to the tales of abundant natural resources in Virginia is contained in this letter from one Tho. Niccolls to Sir Jo. Worstenholme in London in 1623:

If the Company would allow to each man a pound of butter and a portion of cheese weekly, they would find more comfort therein then by all the deer, fish, and fowl [that] is so talked of in England, of which, I can assure you, your poor servants have not had so much as the scent since their coming into the country.

To prevent profiteering in Canadian fish the Virginia authorities had set the selling prices:

January 3, 1625-6: Proclamation by the Governor and Council of Virginia renewing a former proclamation of August 31, 1623, restraining the excessive rates of commodities—commanding that no person in Virginia, either adventurer or planter, shall vend, utter, barter, or sell any of the commodities following above the prices hereafter mentioned, viz: New Foundland fish, the hundred ... 10 pounds of tobacco; Canada dry fish, the hundred ... 24 pounds of tobacco; Canada wet fish, the hundred.... 30 pounds of tobacco.

In one proposed deal of fish for tobacco the owner of the fish got scared off, as recorded in the Minutes of the Council and General Court, 1622-29:

Luke Edan, sworn and examined, says that there were sixteen thousand fish offered him by one Corbin at Canada which afterward the said Corbin refused to sell him for it was told him his tobacco was not good, and as the examiner heard, it was Henry Hewat that told him so.

A case of special concession for the sale of fish was shown in a ruling of the Virginia Council in 1626:

It is ordered that whereas Mr. Weston came up to James City, he shall sell 3,000 of his fish there, which he has promised to sell at reasonable rates. Therefore, in regard the proclamations are not published for the choosing of merchants and factors, it is permitted that such as are desirous to buy any of the said fish he may have leave to deal with Mr. Weston, notwithstanding orders to the contrary.

Another dissuading factor in the unsubstantial fishing in Virginia was the threat of Indian attack. The Assembly in 1626 ruled:

It is ordered, according to the act of the late General Assembly, that no man go or send abroad either upon fowling, fishing, or otherwise whatsoever without a sufficient plenty of men, well armed and provided of munition, upon penalty of undergoing severe censure of punishment by the Governor and Council.

It was characteristic of Virginia's fisheries that the pessimists occupied the stage for a while, then the optimists. An example of the whipping-up of enthusiasm is this discourse of Edward Williams writing on Virginia at mid-century. China was a fabulous country, therefore he compared Virginia with it. Ideas ran riot as he contemplated the resources crying to be developed:

... What multitudes of fish to satisfy the most voluptuous of wishes, can China glory in which Virginia may not in justice boast of?... Let her publish a precedent so worthy of admiration (and which will not admit belief in those bosoms where the eye cannot be witness of the action) of five thousand fish taken at one draught near Cape Charles, at the entry into Chesapeake bay, and which swells the wonder greater, not one fish under the measure of two feet in length. What fleets come yearly upon the coasts of Newfoundland and New England for fish, with an incredible return? Yet it is a most assured truth that if they would make experiment upon the south of Cape Cod, and from thence to the coast of this happy country, they would find fish of greater delicacy, and as full handed plenty, which though foreigners know not, yet if our own planters would make use of it, would yield them a revenue which cannot admit of any diminution while there are ebbs and floods, rivers feed and receive the ocean, or nature fails in (the elemental original of all things) waters.

There wants nothing but industrious spirits and encouragement to make a rich staple of this commodity; and would the Virginians but make salt pits, in which they have a greater convenience of tides (that part of the universe by reason of a full influence of the moon upon the almost limitless Atlantic causing the most spacious fluxes and refluxes, that any shore of the other divisions in the world is sensible of) to leave their pits full of salt-water, and more friendly and warm sunbeams to concoct it into salt, than Rochel, or any parts of Europe. Yet notwithstanding these advantages which prefer Virginia before Rochel, the French king raises a large proportion of his revenues out of that staple yearly, with which he supplies a great part of Christendom.

Nor would it be such a long interval (salt being first made) betwixt the undertaking of this fishing, and the bringing it to perfection, for if every servant were enjoined to practice rowing, to be taught to handle sails, and trim a vessel, a work easily practised, and suddenly learned, the pleasantness of weather in fishing season, the delicacy of the fish, of which they usually feed themselves with the best, the encouragement of some share in the profit, and their understanding what their own benefit may be when their freedom gives them an equality, will make them willing and able fishermen and seamen. To add further to this, if we consider the abundance, largeness, and peculiar excellency of the sturgeon in that country, it will not fall into the least of scruples, but that one species will be of an invaluable profit to the buyer, or if we repeat to our thoughts the singular plenty of herrings and mackerel, in goodness and greatness much exceeding whatever of that kind these our seas produce, a very ordinary understanding may at the first inspection perceive that it will be no great difficulty to out-labor and out-vie the Hollander in that his almost only staple.

This flowery author goes on to make ingenious suggestions about raising fish in captivity, like domesticated animals, by inclosing a creek against their egress but keeping it sluiced to permit the action of tides. He even guesses that a nutritious and medicinal oil could be produced from fish livers. It is worth noting that both these suggestions have been proved practical but they had to wait until modern times to be carried out.

In the anonymous *A Perfect Description of Virginia*, published in 1649, the population is given as 15,000 English and 300 negroes. The count of boats, remembering the shortage of 40 years before, is impressive: "They have in their Colony pinnaces, barks, great and small boats many hundreds, for most of their plantations stand upon the river sides or up little creeks, and but a small way into the land so that for transportation and fishing they use many boats."

The enmity of the Indians had been a constant irritation, and worse, ever since the first days. As soon as it became possible to do so, effort was made to cut them off from the resources of the tidal waters. It was reasoned, and as it turned out, rightly, that with them unable to supplement their food supplies with fish and shellfish, especially oysters, they would be weakened in body and more easily subdued. The word early went out: Keep the Indians away from the water. This strategy worked so successfully that by 1662 it was deemed safe to ease the pressure. Thus another milestone was reached: the first oyster licensing law, as recorded in Hening's *Statutes*:

Be it further enacted that for the better relief of the poor Indians whom the seating of the English had forced from their wonted convenience of oystering, fishing ... that the said Indians upon address made to two of the justices of that county they desire to oyster ... they, the said justices, shall grant a license to the said Indians to oyster ... provided the said justices limit the time the Indians are to stay, and the Indians bring not with them any guns, or ammunition or any other offensive weapon but only such tools or implements as serve for the end of their coming. If any Englishman shall presume to take from the Indians so coming in any of their goods, or shall kill, wound, maim any Indian, he shall suffer as he had done the same to an Englishman and be fined for his contempt.

This was followed, according to Hening, in 1676 by another cavalier gesture to the oppressed:

... It is hereby intended that our neighbor Indian friends be not debarred from fishing and hunting within their own limits and bounds, using bows and arrows only. Provided also that such neighbor Indian friends who have occasion for corn to relieve their lives and it shall and may be lawful for any English to employ in fishing or deal with fish, canoes, bowls, mats, or baskets, and to pay the said Indians for the same in Indian corn, but no other commodities....

Thomas Glover, author of *An Account of Virginia*, addressed to the Royal Society in London, published in 1676, sides with the optimists. His catalogue has a familiar sound but it is valuable as substantiating many of the earlier reports. One impression to be gained from it is that after more than 60 years of occupancy of the new territory, the settlers had in no way depleted their fishery resources, had not, in fact, even scratched the surface:

In the rivers are great plenty and variety of delicate fish. One kind whereof is by the English called a sheepshead from the resemblance the eye of it bears with the eye of a sheep. This fish is generally about fifteen or sixteen inches long and about half a foot broad. It is a wholesome and pleasant fish and of easy digestion. A planter does often times take a dozen or fourteen in an hour's time with hook and line.

There is another sort which the English call a drum, many of which are two foot and a half or three foot long. This is likewise a very good fish, and there is plenty of them. In the head of this fish there is a jelly, which being taken and dried in the sun, then beaten to powder and given in broth, procures speedy delivery to women in labour.

At the heads of the rivers there are sturgeon and in the creeks are great store of small fish, as perch, croakers, taylors, eels, and divers others whose name I know not. Here are such plenty of oysters as they may load ships with them. At the mouth of Elizabeth River, when it is low water, they appear in rocks a foot above water. There are also in some places great store of mussels and cockles. There is also a fish called a stingray, which resembles a skate, only on one side of his tail grows out a sharp bone like a bodkin about four or five inches long, with which he sticks and wounds other fish and then preys upon them.

The same author went farther than any other reporter up to that time in telling a real fish story:

And now it comes into my mind, I shall here insert an account of a very strange fish or rather a monster, which I happened to see in

Rappahannock River about a year before I came out of the country; the manner of it was thus:

As I was coming down the forementioned river in a sloop bound for the bay, it happened to prove calm, at which time we were three leagues short of the river's mouth; the tide of ebb being then done, the sloop-man dropped his grapline, and he and his boy took a little boat belonging to the sloop, in which they went ashore for water, leaving me aboard alone, in which time I took a small book out of my pocket and sat down at the stern of the vessel to read; but I had not read long before I heard a great rushing and flashing of the water, which caused me suddenly to look up, and about half a stone's cast from me appeared a prodigious creature, much resembling a man, only somewhat larger, standing right up in the water with his head, neck, shoulders, breast and waist, to the cubits of his arms, above water; his skin was tawny, much like that of an Indian; the figure of his head was pyramidal, and slick, without hair; his eyes large and black, and so were his eyebrows; his mouth very wide, with a broad streak on the upper lip, which turned upward at each end like mustachioes; his countenance was grim and terrible; his neck, shoulders, arms, breast and waist were like unto the neck, arms, shoulders, breast and waist of a man; his hands if he had any, were under water; he seemed to stand with his eyes fixed on me for some time, and afterward dived down, and a little after riseth at somewhat a farther distance, and turned his head towards me again, and then immediately falleth a little under water, and swimmeth away so near the top of the water, that I could discern him throw out his arms, and gather them in as a man doth when he swimmeth. At last he shoots with his head downwards, by which means he cast his tail above the water, which exactly resembled the tail of a fish with a broad fane at the end of it.

Judging from the few piddling regulations and restrictions referred to in extracts already cited, the Virginia lawmakers could see no need for intensive or even active supervision of the Tidewater fisheries. A rather epoch-making law was enacted in 1678 by the county court of Middlesex County, which is about 50 miles from James City, at the juncture of the Rappahannock river and Chesapeake bay:

Whereas, by the 15th act of Assembly made in the year 1662, liberty is given to each respective county to make by-laws for themselves; which laws, by virtue of the said act are to be binding upon them as any other general law; and whereas several of the inhabitants of this county have complained against the excessive and immoderate striking and destroying of fish, by some fire, of the inhabitants of this county by striking them by a light in the night time with fish gigs, wherby they not only affright the fish from coming into the rivers and creeks, but also wound four times that quantity that they take, so that if a timely remedy be not applied, by that means the fishing with hooks and lines will be thereby spoiled to the great hurt and grievance of most of the inhabitants of this county. It is therefore by this court ordered that from and after the 20th day of March next ensuing, it shall not be lawful for any of the inhabitants of this county to take, strike, or destroy any sort of fish in the night time with fish gigs, harping irons, or any other instrument of that nature, sort or kind, within any river, creek or bay which are accounted belonging to or within the bounds or precincts of this county. And it is further ordered that if any person or persons being a freeman, shall offend against this order, he or they so offending shall for the first offence be fined five hundred pounds of good tobacco to be paid to the informer, and for every other offence committed against this order after the first, by any person, the said fine to be doubled and if any servants be permitted or encouraged by their masters to keep or have in their possession any fish gig, harping iron or any other instrument of that kind or nature and shall therewith offend against this order, that in such case the master of such servant or servants shall be liable to pay the several fines above mentioned, and if any servant or servants shall, contrary to and against their master's will and knowledge, offend against this order, that for every offence they receive such corporal punishment as by this court shall be thought meet.

As population became more dense it was inevitable that rights previously of little significance began to be asserted. This case of 1679 taken from Hening's *Statutes*, was a forerunner of countless others like it which continue to this day:

Robert Liny, having complained to this Grand Assembly that whereas he had cleared a fishing place in the river against his own land to his great cost and charge supposing the right thereof in himself by virtue of his patents, yet nevertheless several persons have frequently obstructed him in his just privilege of fishing there, and despite of him came upon his land and hauled their seines on shore to his great prejudice, alleging that the water was the King Majesty's and not by him granted away in any patent and therefore equally free to all His Majesty's subjects to fish in and haul their seines on shore, and praying for relief therein by a declaratory order of this Grand Assembly; it is ordered and declared by this Grand Assembly that every man's right by virtue of his patent extends into the rivers or creeks so far as low water mark and it is a privilege granted to him in and by his patent, and that therefore no person ought to come and fish there above low water mark or haul seines on shore without leave first obtained, under the hazard of comitting a trespass for which he is sueable by law.

In most cases this decision somewhat limited a landowner's claim. But on the seaside of Virginia's Eastern Shore conditions have always been so that at low tide thousands of acres of land are laid bare, with the result that "low water mark" is in many cases difficult of interpretation as a boundary between waterfront properties and the public domain.

Toward the close of the century fishing methods had shaped up advantageously compared to the crudities and hit-or-miss practices of the first settlers. Robert Beverley described them in 1705:

The Indian invention of weirs in fishing is mightily improved by the English, besides which, they make use of seines, trolls, casting nets, setting nets, hand fishing and angling and in each find abundance of diversion. I have sat in the shade at the heads of the rivers angling and spent as much time in taking the fish off the hook as in waiting for their taking it. Like those of the Euxine Sea, they also fish with spilyards which is a long line staked out in the river and hung with a great many hooks on short strings, fastened to the main line, about four foot asunder. The only difference is that our line is supported by stakes and theirs is buoyed up with gourds.

The abundance of the fisheries never ceased impressing visitors. A French tourist added to the chorus in 1687:

Fish too is wonderfully plentiful. There are so many shell oysters that almost every Saturday my host craved them. He had only to send one of his servants in one of the small boats and two hours after ebb tide he brought it back full. These boats, made of a single tree hollowed in the middle, can hold as many as fourteen people and twenty-five hundredweight of merchandise.

As if to crown the final emergence of recognition of the home fisheries William Byrd I instructed his agent in Boston in 1689 to send him a variety of commodities in return for a bill of exchange but *no salt fish*:

By the advice of my friend, Captain Peter Perry, I made bold to give you the trouble of a letter of the 1st instant with two small bills of exchange which I desired you to receive and return the effects to me in the upper part of James River, either in rum, sugar, Madeira wine, turnery, earthenware, or anything else you may judge convenient to this country (fish excepted)....

Evidently at least some good salt was now at hand to preserve the roe herring that choked the rivers and creeks in the spring. The salt-herring breakfast was on its way to becoming a Virginia institution, and the salt-fish monopolies of New England and Canada were cracking after three-quarters of a century.

The score of "firsts" in the Virginia fishery world have been noted as they occurred. Among them were the first fishery statistics, the first licensing law, the first price control, the first diamond-back terrapin, the first conservation measures. And now in 1698 there was the first agitation against polluted waters:

We, the Council and Burgesses of the present General Assembly, being sensible to the great mischiefs and inconveniences that accrue to the inhabitants of this, his Majesty's Colony and Dominion of Virginia, by killing of whales within the capes thereof, in all humility take leave to represent the same unto Your Excellency and withal to acquaint you that by the means thereof great quantities of fish are poisoned and destroyed and the rivers also made noisome and offensive. For prevention of which evils in regard the restraint of the killing of whales is a branch of His Majesty's royal prerogative.

We humbly pray that Your Excellency [the Governor, Francis Nicholson] will be pleased to issue out a proclamation forbidding all persons whatsoever to strike or kill any whales within the bay of Chesapeake in the limits of Virginia which we hope will prove an effectual means to prevent the many evils that arise therefrom.

As Jamestown reached the end of its span, the fisheries came of age. Inequities were being ironed out, methods were being perfected, and planners were at work on ways of employing more and more of the fast-growing population in searching out and making available the bounty of the fair Chesapeake.

At the start of the 18th century, however, there was little evidence of an organized industry in any phase. Everywhere were unlimited opportunities for exploitation. The abundance of oysters still impressed travelers. In the extract to follow, Francis Louis Michel of Switzerland speaks of the method of tonging oysters in 1701, but note that he says, "They usually pull from six to ten times." This could be taken to mean that each individual procured his own oysters from the lavish supply virtually at his doorstep, and stopped as soon as he had a "mess" to enjoy over the week-end:

The water is no less prolific, because an indescribably large number of big and little fish are found in the many creeks, as well as in the large rivers. The abundance is so great and they are so easily caught that I was much surprised. Many fish are dried, especially those that are fat. Those who have a line can catch as many as they please. Most of them are caught with the hook or the spear, as I know from personal experience, for when I went out several times with the line, I was surprised that I could pull out one fish after another, and, through the clear water I could see a large number of all kinds, whose names are unknown to me. They cannot be compared with our fish, except the herring, which is caught and dried in large numbers. Thus the so-called catfish is not unlike the large turbot. A very good fish and one easily caught is the eel, also like those here [in Switzerland]. There is also a kind like a pike. They have a long and pointed mouth, with which they like to bite into the hook. They are not wild, but it happens rarely that one can keep them on the line, for they cut it in two with their sharp teeth. We always had our harpoons and guns with us when we went out fishing, and when the fish came near we shot at them or harpooned them. A good fish, which is common and found in large numbers is the porpoise. They are so large that by their unusual leaps, especially when the weather changes, they make a great noise and often cause anxiety for the small boats or canoes. Especially do they endanger those that bathe. Once I cooled and amused myself in the water with swimming, not knowing that there was any danger, but my host informed me that there was.... The waters and especially the tributaries are filled with turtles. They show themselves in large numbers when it is warm. Then they come to the land or climb up on pieces of wood or trees lying in the water. When one travels in a ship their heads can be seen everywhere coming out of the water. The abundance of oysters is incredible. There are whole banks of them so that the ships must avoid them. A sloop, which was to land us at Kingscreek, struck an oyster bed, where we had to wait about two hours for the tide. They surpass those in England by far in size, indeed, they are four times as large. I often cut them in two, before I could put them into my mouth. The inhabitants usually catch them on Saturday. It is not troublesome. A pair of wooden tongs is needed. Below they are wide, tipped with iron. At the time of the ebb they row to the beds and with the long tongs they reach down to the bottom. They pinch them together tightly and then pull or tear up that which has been seized. They usually pull from six to ten times. In summer they are not very good, but unhealthy and can cause fever.

The most comprehensive list of fish thus far given by the early historians was offered by Robert Beverley in 1705. Again as with John Smith, there are names that do not fit in today. But these are very few: "greenfish," "maid," "wife," and "frogfish" perhaps, all of which, however, are well-known in England. The recurring mention of carp in the early authorities quoted is interesting, since it has long been believed that carp were introduced into the Chesapeake region in 1877 by the U.S. Fish Commission. No doubt that was carp of another species. The esteemed sheepshead is today very rare:

As for fish, both of fresh and salt water, of shellfish, and others, no country can boast of more variety, greater plenty, or of better in their several kinds.

In the spring of the year, herrings come up in such abundance into their brooks and fords to spawn that it is almost impossible to ride through without treading on them. Thus do those poor creatures expose their own lives to some hazard out of their care to find a more convenient reception for their young, which are not yet alive. Thence it is that at this time of the year, the freshes of the rivers, like that of the Broadruck, stink of fish.

Besides these herrings, there come up likewise into the freshes from the sea multitudes of shad, rock, sturgeon, and some few lampreys, which fasten themselves to the shad, as the remora of Imperatus is said to do to the shark of Tiburon. They continue their stay there about three months. The shad at their first coming up are fat and fleshy, but they waste so extremely in milting and spawning that at their going down they are poor and seem fuller of bones, only because they have less flesh. As these are in the freshes, so the salts afford at certain times of the year many other kinds of fish in infinite shoals, such as the oldwife, a fish not much unlike a herring, and the sheepshead, a sort of fish which they esteem in the number of their best.

There is likewise great plenty of other fish all the summer long and almost in every part of the rivers and brooks there are found of different kinds. Wherefore I shall not pretend to give a detail of them, but venture to mention the names only of such as I have eaten and seen myself and so leave the rest to those that are better skilled in natural history. However, I may add that besides all those that I have met with myself, I have heard of a great many very good sorts, both in the salts and freshes, and such people too, as have not always spent their time in that country, have commended them to me, beyond any they had ever eaten before.

Those which I know myself, I remember by the names of herring, rock, sturgeon, shad, oldwife, sheepshead, black and red drums, trout, taylor, greenfish, sunfish, bass, chub, plaice, flounder, whiting, fatback, maid, wife, small turtle, crab, oyster, mussel, cockle, shrimp, needlefish, bream, carp, pike, jack, mullet, eel, conger eel, perch, and catfish.

Those which I remember to have seen there of the kinds that are not eaten are the whale, porpoise, shark, dogfish, gar, stingray, thornback, sawfish, toadfish, frogfish, land crabs, fiddlers, and periwinkle.

Francis Makemie, often called the father of American Presbyterianism, was concerned, in his *A Plain and Friendly Perswasive to the Inhabitants of Virginia and Maryland for Promoting Towns and Cohabitations*, about the dearth of markets for fishery products. It was a condition brought about largely by a general lack of money in circulation. It was easily possible for entire families to subsist the year around on the fruits of land and water plus unexacting manual labor. Wealth was concentrated in the hands of the more important planters whose estates were usually self-sufficient and concentrating on trade with England. The natural bounty of the Tidewater region thus actually deterred the development of Virginia along the lines of New England with its urban centers:

Cohabitation would not only employ thousands of people ... others would be employed in hunting, fishing, and fowling, and the more diligently if assured of a public market....

So also our fishing would be advanced and improved highly by encouraging many poor men to follow that calling, and sundry sorts which are now slighted would be fit for a town market, as sturgeon, thornback, and catfish. Our vast plenty of oysters would make a beneficial trade, both with the town and foreign traders, believing we have the best oysters for pickling and transportation if carefully and skillfully managed.

By 1705 the seat of government had been transferred to nearby Williamsburg. The need of establishing towns as foci for the developing countryside had been felt and now the legislators turned their attention to promoting the fish markets therein, followed by some essential protection of the rights of fishermen and others. Hening's *Statutes* gives the details:

October, 1705. For the encouragement and bettering of the markets in the said town, Be it enacted, That no dead provision, either of flesh or fish shall be sold within five miles of any of the ports or towns appointed by this act, on the same side the great river the town shall stand upon, but within the limits of the town, on pain of forfeiture and loss of all such provision by the purchases, and the purchase money of such provision sold by the vendor, cognizable by any justice of the county....

Be it further enacted and declared, That if any person or persons shall at any time hereafter shoot, hunt or range upon the lands and tenements, or fish or fowl in any creeks or waters included within the lands of any other person or persons without license for the same, first obtained of the owner and proprietor thereof, every such person so shooting, hunting, fishing, fowling, or ranging, shall forfeit and pay for every such offence, the sum of five hundred pounds of tobacco....

Be it further enacted, That if any person shall set, or cause to be set, a weir in any river or creek, such person shall cause the stayes thereof to be taken up again, as soon as the weir becomes useless; and if any person shall fail of performing his duty herein, he shall forfeit and pay fifteen shillings current money, to the informer: To be recovered, with costs, before a justice of the peace.

The essentials of any stable industry are: control of supply and means of distribution. The fisheries of Virginia were blessed with neither of these advantages. Any progress had to be made in spite of uncertain harvests and lack of packing and handling facilities. Distribution of fresh seafoods was impossible without rapid transportation and adequate refrigeration. Neither was available for two centuries. Virginia's huge supply of oysters was a case in point. Consumption of oysters was limited to those who lived on the spot, and though they figured importantly in the Tidewater diet, as a palpable resource they were untouched until the 19th century. The principal means of preserving them before then was by pickling. In that form they were quite popular during the Colonial period. Fish were salted when there was a surplus and in certain seasons, especially the spawning time of the anadromous river-herring, they were available in phenomenal quantities. They remain today among Virginia's most plentiful fish but the salting industry has now become a mere token of its former magnitude.

The Chesapeake bay blue crab which today constitutes a resource worth about $5,000,000 a year to Virginia crabbers and packers, had to wait even longer than fish and oysters did for development. Salting and pickling were unsuitable to this delicate food and expeditious handling methods did not exist.

In an exhaustive catalogue of the marine life of Virginia William Byrd II, of Westover said:

Herring are not as large as the European ones, but better and more delicious. After being salted they become red. If one prepares them with vinegar and olive oil, they then taste like anchovies or sardines, since they are far better in salt than the English or European herring. When they spawn, all streams and waters are completely filled with them, and one might believe, when he sees such terrible amounts of them, that there was as great a supply of herring as there is water. In a word, it is unbelievable, indeed, indescribable, as also incomprehensible, what quantity is found there. One must behold oneself.

At the time he wrote Virginians were beginning to compete with Canadians and New Englanders in exporting salt fish, particularly to the West Indies, where a large proportion of them were exchanged for the rum so freely used on the plantations as slave rations.

There were no dams barring access to the highest reaches of the rivers and no cities and factories to discharge pollution, so that the river-herring and shad made their way far inland even to the Blue Ridge mountains. There the pioneers awaited them eagerly each spring and salted down a supply to tide them over till the next run. Small wonder, then, that the love of salt herring—always with corn bread—became ingrained in so many Old Virginians!

They had an illustrious exemplar. Once, in 1782, when George Washington was due to visit Robert Howe the honored host wrote to a friend: "General Washington dines with me tomorrow. He is exceedingly fond of salt fish."

Despite obstacles a healthy experimentation in the various phases of fishing was now and then manifest. For example, in 1710 one adventurous fisherman wished to extend the home fisheries to whaling and applied to the Virginia Council for a license. Whales, though not common in Chesapeake bay or the ocean area near it, had been noted from time to time ever since the birth of the Colony.

Most often they were washed ashore dead. John Custis, of Northampton County, succeeded in making 30 barrels of oil from one such in 1747. The year before that a live one was spotted in the James river by some Scottish sailors who were able to comer it in shallow water. After killing it, they found it to measure 54 feet! The *Virginia Gazette*, published in Williamsburg, carried this item in 1751:

Some principal gentlemen of the Colony, having by voluntary subscription agreed to fit out vessels to be employed in the whale fishery on our coast, a small sloop called the *Experiment* was some time ago sent on a cruise, and we have the pleasure to acquaint the public that she is now returned with a valuable whale. Though she is the first vessel sent from Virginia in this employ, yet her success, we hope, will give encouragement to the further prosecution of the design which, we doubt not, will tend very much to the advantage of the Colony as well as excite us to other profitable undertakings hitherto too much neglected.

Commented John Blair in his *Diary* on the incident: "Heard our first whale brought in and three more struck but lost." The *Experiment* continued its whaling career successfully for three years. When it retired, no similar enterprise replaced it. Yet in a list of exports from Virginia for the year ending September 30, 1791, 1263 gallons of whale oil appears. Even today whales are occasionally represented in Virginia fishery products, as when one is washed up on a beach and removed by the Coast Guard to a processing plant to be turned into meal and oil.

The overall value of Virginia's fisheries as an industrial resource was glacially slow in reaching public consciousness. Here and there, like dim lights along an uncertain voyage, bits of legislation or isolated conservation procedures appeared. In due course it became evident that natural fishways—to choose one example—were being obstructed to the disadvantage of both the fish and navigation. Hening records the law enacted to keep the rivers open:

1745. And whereas the making and raising of mill dams, and stone-stops, or hedges for catching of fish, is a great obstruction to the navigation of the said rivers [James and Appomattox]: Be it further enacted by the authority aforesaid, That all mill dams, stone-stops, and hedges, already made across either of the said rivers, where they are navigable, shall be thrown down and destroyed by the person or persons who made the same....

Like most hastily framed and passed laws this one proved unsatisfactory and a second one, with more detailed provisions was passed. Hening records it:

1762. Whereas the act of assembly made in the first year of his present Majesty's reign [1761], entitled, an act to oblige the owners of mills, hedges, or stone-stops, on sundry rivers therein mentioned, to make openings or slopes therein for the passage of fish, has been found defective, and not to answer the purposes for which it was intended, and it is therefore necessary that the same should be amended: Be it therefore enacted by the Lieutenant Governor, Council and Burgesses, of this present General Assembly, and it is hereby enacted by the authority of the same, That the owner or proprietor of all and every mill, hedge, or stone-stop, on either of the rivers Nottoway and Meherrin, shall in the space of nine months from and after the passing of this act, make an opening or slope in their respective mill-dams, hedges, or stops, in that part of the same where there shall happen to be the deepest water, which shall be in width at least ten feet in the clear, in length at least three times the height of the dam, and that the bottoms and sides thereof shall be planked, and that the sides shall be at least fourteen inches deep, so as to admit a current of water through the same twelve inches deep, which shall be kept open from the tenth day of February to the last day of May in every year.... And be it further enacted by the authority aforesaid, That if any such owner or proprietor shall neglect or refuse so to do, within the time aforesaid, the person so offending shall forfeit and pay the sum of five pounds of tobacco for every day he or they shall so neglect or refuse....

Still the fundamental problem was not solved; fish were not bypassing the remaining obstructions in sufficient quantity to maintain the expected harvest. After various amendments and additions this explicit definition of a fishway or slope was enacted into law in 1771:

That a gap be cut in the top of the dam contiguous to the deepest part of the water below the dam, in which shall be set a slope ten feet wide, and so deep that the water may run through it 18 inches before it will through the waste, or over the dam, that the direction of the said slope be so, as with a perpendicular to be dropped from the top of the dam, will form an angle of at least 75 degrees, and to continue in that direction to the bottom of the river, below the dam, to be planked up the sides 2 feet high; that there be pits or basins built in the bottom, at 8 feet distance, the width of the said slope, and to be 12 inches deep, and that the whole be tight and strong; which said slope shall be kept open from the 10th day of February to the last day of May, annually, and any owner not complying to forfeit 5 pounds of tobacco a day.

The effort was of little avail. Before many dams could be so laboriously modified the Revolutionary War arrived to obscure placid matters like fish conservation.

The diaries of the 18th Century Virginia planters abound with references to seafoods. Most of them lived either on or within easy distance of Tidewater. Most of them had nets and other fishing implements of their own and crews among the slaves to work them. Whenever their needs required, an expedition was made. Perhaps there was a season of bountiful entertaining in prospect. The seine would be taken to a likely spot and hauled ashore. Or a boat would go out and load up with oysters. The fish had to be eaten right away or salted down. But oysters stored in a dark cellar, especially in cool weather, would keep for weeks if moistened from time to time.

One diarist, James Gordon, lived near the Rappahannock river in a section affording a variety of seafoods. Note these typical entries:

Sept. 20, 1759. Fine weather. Went in the afternoon and drew the seine. Had very agreeable diversion and got great plenty of fine fish....

Sept. 26. Went with my wife in the evening to draw the seine. Got about sixty greenfish and a few other sorts.

Sept. 28. Sent in the morning to have the seine drawn. They made several hauls and got good fish, viz: three drum, one of them large, trouts, greenfish, etc....

Oct. 6. Went with my wife to see the seine drawn. We dined very agreeably on a point on fish and oysters....

Jan. 22,—Bought about 70 gallons of rum. Got fine oysters there.

Feb. 12. Went on board the New England man and bought some pots, axes and mackerel.

Feb. 22. Drew the seine and got 125 fine rock and some shad.

July 14. Drew the seine today and got some fine rock.

Feb. 9, 1760. Went with my wife and Mr. Criswell to draw the seine. We met in Eyck's Creek a school of rock—brought up 260. Some very large; the finest haul I ever saw. Sent many of them to our neighbors.

The term "greenfish" is unknown among Virginia Tidewater fishermen. Here again we have a British name brought into Virginia by a colonist not long removed from that country. There "greenfish" is applied to the bluefish, of which there were and are at times plenty in the Rappahannock river.

Another diarist, who lived only a few miles away from Gordon, also on the Rappahannock river, was Landon Carter, son of the famed Robert, or "King," Carter of Corotoman in Lancaster County. There is no doubt about it: he was an oyster lover. He not only knew a way to hold oysters over an extended period—one wishes one knew what it was—but he had the courage and originality to eat them in July, contrary to a widely respected superstition:

Jan. 14, 1770. My annual entertainment began on Monday, the 8th, and held till Wednesday night, when, except one individual or two that retired sooner, things pleased me much, and therefore, I will conclude they gave the same satisfaction to others.

The oysters lasted till the third day of the feast, which to be sure, proves that the methods of keeping them is good, although much disputed by others.

July, 1776. Last night my cart came up from John E. Beale for iron pots to make salt out of the bay water, which cart brought me eight bushels oysters. I ordered them for family and immediate use. As we are obliged to wash the salt we had of Col. Tayloe, I have ordered that washing be carried into the vault and every oyster dipped into it over all and then laid down on the floor again.... Out of the eight bushels oysters I had six pickled and two bushels for dressing. But I was asked why Beale sent oysters up in July. I answered it was my orders. Who would eat oysters in July said the mighty man; and the very day showed he not only could eat them but did it in every shape, raw, stewed, caked in fritters and pickled.

George Washington, too, was an oyster fancier as this note to his New York friend George Taylor shows:

Mt. Vernon, 1786. Sir: ... Mrs. Washington joins me in thanking you also for your kind present of pickled oysters which were very fine. This mark of your politeness is flattering and we beg you to accept every good wish of ours in return.

When in 1770 a notice appeared in the *Virginia Gazette* about the proposed academy in New Kent County an added attraction was featured: "Among other things the fine fishery at the place will admit of an agreeable and salutary exercise and amusement all the year." It was the Chickahominy river, a tributary of the James, that was referred to. Fishing is still "agreeable" there. Citizens of Richmond, recreation-bent, throng to it along with the residents of its banks, many of whom make their living out of it. This is one of the sections where the water, though tidal, is fresh. Anadromous herring, shad, rock and sturgeon are caught. Unlike the salty bay, fish can be caught here the year round. Among them are catfish, carp, perch and bass.

One of the most accurate and vivid reporters of Colonial Virginia plantation life was Philip Vickers Fithian, tutor to the family of Councillor Robert Carter of Nominy Hall on the lower Potomac river. In his *Journals* are appetizing references to seafood:

1774, March: With Mr. Randolph, I went a-fishing, but we had only the luck to catch one apiece.

April. We had an elegant dinner; beef and greens, roast pig, fine boiled rockfish.

July. We dined today on the fish called the sheepshead, with crabs. Twice every week we have fine fish.

On the edges of these shoals in Nominy River or in holes between the rocks is plenty of fish.

Well, Ben, you and Mr. Fithian are invited by Mr. Turberville, to a fish feast tomorrow, said Mr. Carter when we entered the Hall to dinner.

As we were rowing up Nominy we saw fishermen in great numbers in canoes and almost constantly taking in fish,—bass and perch.

This is a fine sheepshead, Mr. Stadly [the music master], shall I help you? Or would you prefer a bass or a perch? Or perhaps you will rather help yourself to some picked crab. It is all extremely fine, sir, I'll help myself.

August. Each Wednesday and Saturday, we dine on fish all the summer, always plenty of rock, perch, and crabs, and often sheepshead and trout.

September. We dined on fish and crabs, which were provided for our company, tomorrow being fish day.

September. Dined on fish,—rock, perch, fine crabs, and a large fresh mackerel.

I was invited this morning by Captain Tibbs to a barbecue. This differs but little from the fish feasts, instead of fish the dinner is roasted pig, with the proper appendages, but the diversion and exercise are the very same at both.

An English traveler in 1759, Andrew Burnaby, registered his wonder at the way fish were taken in the reaches of the Chesapeake:

Sturgeon and shad are in such prodigious numbers [in Chesapeake Bay] that one day within the space of two miles only, some gentlemen in canoes caught above six hundred of the former with hooks, which they let down to the bottom and drew up at a venture when they perceived them to rub against a fish; and of the latter above five thousand have been caught at one single haul of the seine.

The "gentlemen" concerned were obviously not slaves serving the needs of a plantation, but, judging from the amount caught, expert commercial fishermen. The sturgeon, after the roe was removed, were stacked in carts and peddled in nearby towns. The shad, after as many as possible were sold fresh, were salted down.

The snagging of big sturgeon as recounted by the French traveler François J. de Chastellux in 1781 remained in common practice into the 20th Century, when the big ones became much scarcer:

As I was walking by the river side [James near Westover], I saw two negroes carrying an immense sturgeon, and on asking them how they had taken it, they told me that at this season they were so common as to be taken easily in a seine and that fifteen or twenty were found sometimes in the net; but that there was a much more simple method of taking them, which they had just been using. This species of monster, which are so active in the evening as to be perpetually leaping to a great height above the surface of the water, usually sleep profoundly at mid-day. Two or three negroes then proceed in a little boat, furnished with a long cord at the end of which is a sharp iron crook, which they hold suspended like a log line. As soon as they find this line stopped by some obstacle, they draw it forcibly towards them so as to strike the hook into the sturgeon, which they either drag out of the water, or which, after some struggling and losing all his blood, floats at length upon the surface and is easily taken.

The frequently met-with term, "fishery," in Colonial writings took on a special meaning as the industry developed. It was used in the sense of what the present Virginia lawbook calls a "regularly hauled fishing landing."

This is usually a shore privately owned where the fronting waters have been cleared of obstructions. The owner, or some one permitted by him, operates a long seine at that place by carrying it offshore in boats and hauling it to land. So long as he thus uses the spot "regularly" the law protects him, now as in the past, by making it illegal for any other person to fish with nets within a quarter-mile of "any part of the shore of the owner of any such fishery."

The rights to such a property were, and are, in many cases extremely profitable. George Washington was among the Virginia planters zealously caring for their "fisheries."

Often the privilege of using these was advertised in the newspapers or otherwise for rent for a long or short term. Some owners who did not themselves wish to fish counted on their shores to yield rental. One of these, George William Fairfax, must have expressed himself to Washington on the subject, for the latter wrote him in June, 1774:

... As to your fishery at the Raccoon Branch, I think you will be disappointed there likewise as there is no landing on this side of river that rents for more than one half of what you expect for that, and that on the other side opposite to you (equally good they say) to be had at £15 Maryland currency....

But growing along with this practice was sentiment favoring fishing places open to the general public. When an attempt was made about 1770 to take over certain lands near Cape Henry for private operation, a vigorous protest ensued:

The petition of the subscribers, inhabitants of the county of Princess Anne in behalf of themselves and the other inhabitants of this colony, humbly shows: That the point of land called Cape Henry bounded eastward by the Atlantic Ocean, northwardly by Chesapeake Bay, westwardly and southwardly by part of Lynnhaven River and by a creek called Long Creek and the branches thereof, is chiefly desert banks of sand and unfit for tillage or cultivation and contains several thousand acres.

And that for many years past a common fishery has been carried on by many of the inhabitants of said county and others on the shore of the ocean and bay aforesaid, as far as the western mouth of Lynnhaven River. And that during the fishing season the fishermen usually encamp amongst the said sand hills and get wood for fuel and stages from the desert, and that very considerable quantities of fish are annually taken by such fishery which greatly contributes to the support and maintenance of your petitioners and their families.

Your petitioners further show that they have been informed that several gentlemen have petitioned your Honour to have the land aforesaid granted to them by patent and that one Keeling has lately surveyed a part thereof situated near the mouth of Long Creek aforesaid, and that if a patent should be granted for the same, it would greatly prejudice the said fishery.

Your petitioners therefore humbly pray that no patent may be granted to any person or persons for the same lands or any part thereof; and that the same may remain a common for the benefit of the inhabitants of this Colony in general for carrying on a fishery and for such public uses as the same premises shall be found convenient.

Even when the new United States Government erected a lighthouse at Cape Henry a careful stipulation was made in the act ceding the property in 1790 that the public were not to be denied fishing privileges there:

Deed of cession of two acres of land at Cape Henry, in Princess Anne County, Virginia, for the purpose of erecting a lighthouse thereon ... provided that nothing contained in this act shall affect the right of this State to any materials heretofore placed at or near Cape Henry for the purpose of erecting a lighthouse, nor shall the citizens be debarred, in consequence of this cession, from the privileges they now enjoy of hauling their seines and fishing on the shores of the said land so ceded to the United States.

When George Washington had come, a newlywed, to be master of Mt. Vernon in 1759 he found the prospects for fishing very satisfying. One of his letters at this time boasted:

A river [the Potomac] well-stocked with various kinds of fish at all seasons of the year, and in the spring with shad, herrings, bass, carp, perch, sturgeon, etc., in great abundance. The borders of the estate are washed by more than ten miles of tidewater, the whole shore, in fact, is one entire fishery.

Washington generously ordered his overseer to admit "the honest poor" to fishing privileges at one of his shores, a concession that may have been customary among many landowners.

He was a man who believed in keeping records, and so complete a file of them has now been reassembled at Mt. Vernon that it is possible to follow his career in any phase: officer, business speculator, host, farmer, legislative adviser, and friend. He gave to fishing the painstaking personal attention he gave to all else. As a "fisherman" he directed the manufacture as well as the repair of his nets, and the curing, shipping and marketing of his fish.

It seems obvious that suitable nets were not being manufactured in the desired quantity or variety in America, otherwise he would hardly have bought his in England.

He dealt with Robert Cary and Co., London, in 1771. Here is a typical order:

One seine, seventy-five fathoms long when rigged for hauling; to be ten feet deep in the middle and eight at the ends with meshes fit for the herring fishery. The corks to be two and a half feet asunder; the leads five feet apart; to be made of the best three-strand (small) twine and tanned.

400 fathom of white inch rope for hauling the above seine. 150 fathom of deep sea line.

To get ready for spring fishing he had to prepare as far ahead as July. Even then he was not always sure delivery would be on time:

... The goods you will please to forward by the first vessel for Potomac (which possibly may be Captain Jordan the bearer of this) as there are some articles that will be a good deal wanted, especially the seine, which will be altogether useless to me if I do not get them early in the spring, or in other words I shall sustain a considerable disappointment and loss, if they do not get to hand in time.

He wrote to Bradshaw and Davidson in London in 1772:

That I may have my seine net exactly agreeable to directions this year I give you the trouble of receiving this letter from me to desire that three may be made. One of them eighty fathom long, another seventy, and the third sixty-five fathom, all of them to be twelve feet deep in the middle and to decrease to seven at the ends when rigged and fit for use; to be so close-meshed in the middle as not to suffer the herrings (for which kind of fishery they are intended) to hang in them because, when this is the case it gives us a good deal of trouble at the busy hurrying season to disengage the seine, and often is the means of tearing it. But the meshes may widen as they approach the ends: the corks to be no more than two feet and a half asunder and fixed on flatways that they may swim and bear the seine up better with a float right in the middle to show the approach of the seine with greater certainty in case the corks should sink; the leads to be five feet apart. The seine I had from you last year had two faults, one of which is that of having the meshes too open in the middle; the other of being too strait rigged; to avoid which I wish you to loose at least one-third of the length in hanging these seines;

that is, to let your 80 fathom seine be 120 in the strait measure (before it is hung in the lead and cork lines) and the other two to bear the same proportion, I could wish to have these seines tanned but it is thought the one I had from you last year was injured in the vat, for which reason I leave it to you to have these tanned or not, as you shall judge most expedient ... I would not wish to have them made of thick heavy twine as they are more liable to heat and require great force to work them....

A detailed reply came from James Davidson, a partner in the net company:

London, Sept. 29, 1772. Sir: I had the honour of receiving your letter with instructions concerning your seines. I shall always pay due attention to the contents. I persuade myself you'll say I have fulfilled your instructions given me in these three seines which I heartily hope will be in time for the intended fishery. Am not afraid but they will meet with your approbation and if you should see any alteration wanting if you'll be so obliging as to send a line in the same channel, it shall be attended to with great care. Your order is for the corks to be put on flat ways. I have only put them on the 65 fathom seine for these reasons. We have tried that method before with every other invention for the satisfaction of our fishermen here but they have assured us they really do not bear the net up so well. They are obliged to be tied on so tight that the twine cuts them and are much apter to break and after all in dragging the net they will swim sideways. Now, Sir, you'll readily see the above inconveniences. I have also put six floats in the middle, two together to show the center of the net. Likewise the length of the netting, 120 fathoms for the 80 fathoms, the other two in proportion.

I now enter upon tanning. This, you may assure yourself, they are pretty well wore if you have them tanned for we are obliged to haul them in and out to take the tan and after that hauling them about to get them thoroughly dry before we can possibly pack them or else they would soon rot. Among the hundreds of seines I sent abroad last year or this, I only tanned one besides yours. Therefore have not tanned any of these. I think the three-quarters inch mesh that I have put in the middle of the nets this year will be a cure for the malady you mention of the herrings hanging in the mesh, for last year I only put in inch mesh which upon examination you'll soon perceive. Therefore, sir, I entreat the honour of a line whether or not the two above three-quarters mesh seines answer the purpose. I have tapered them away at the ends to [an] inch and a half.

These nets were designed for hauling ashore by hand. It was not till much later that other nets, of the styles so familiar today, gill nets and pound nets in particular, came into general use.

Much longer seines than Washington needed were used as fish became scarcer. There are tales of them four and five miles long, actually able to block off the entire river, being used in the neighborhood of Mt. Vernon before control laws were enacted and enforced. The catches were enormous. Barges were heaped high with all sorts of fish and towed into Washington City where they were sold before they spoiled, for what they would bring.

Today the pollution for which Washington and Alexandria are responsible has destroyed most fish life within several miles of Mt. Vernon.

Like his fishing predecessors ever since Jamestown, Washington had his troubles with salt. One of his business letters ordering a supply complained: "Liverpool salt is inadequate to the saving of fish.... Lisbon is the proper kind."

He was only briefly touching on a subject that had vexed the Colonists since the beginning. Through the years the cry for more and better salt had gone up. The fishermen of Virginia needed salt for their fish as badly as the Hebrews in Egypt needed straw for their bricks. Although trading with foreign countries increased steadily, the question of a salt supply for Virginia remained unsolved.

As the 18th century had progressed, matters grew even worse. In 1763 the Virginia Committee of Correspondence had written urgently to its agent in London to apply to Parliament for an act to

allow to this Colony the same liberty to import salt from Lisbon or any other European ports, which they have long enjoyed in the Colonies and provinces of New England, New York and Pennsylvania. This is a point that hath been more than once unsuccessfully labored; but we think it is so reasonable, that when it is set in a proper light, we shall hope for success. The reason upon which the opposition hath been supported, is this general one that it is contrary to the interest of Great Britain to permit her plantations to be supplied with any commodity, especially any manufacture from a foreign country, which she herself can supply them with. This we allow to be of force; provided the Mother Country can and does supply her plantations with as much as they want; but the fact being otherwise, we have been allowed to supply ourselves with large quantities from Cercera, Isle of May, Sal Tortuga and so forth. The course of this trade being hazardous, in time of war, this useful and necessary article hath been brought to us at a high price of late. The reason or pretence of granting this indulgence to the Northern Colonies, in exclusion of the Southern, we presume to be to enable them to carry on their fishery to greater advantage, the salt from the Continent of Europe being fitter for that purpose than the salt from Great Britain or that from any of the islands we have mentioned. But surely this reason is but weakly founded with respect to Pennsylvania, whose rivers scarcely supply them with fish sufficient for their own use; whereas the Bay of Chesapeake abounds with great plenty and variety of fish fit for foreign markets, as well as for ourselves, if we could but get the proper kind of salt to cure it. Herrings and shads might be exported to the West Indies to great advantage; and we could supply the British markets with finer sturgeon than they have yet tasted from the Baltic. And it is an allowed principle that every extension of the trade of the Colonies, which does not interfere with that of the Mother Country is an advantage to the latter; since all our profits ultimately center with her.

It was pointed out that the English merchants were not above sharp practices in filling orders for salt; they would reduce the amount shipped to individuals and provide the captain with all he could carry extra to be sold at high prices to needy buyers.

The plaint was just another of the rumblings of discontent contributing to the grand explosion of thirteen years later. The intricacies were entered into in detail by the Committee:

We have twelve different Colonies on the Continent of North America. Four of them, viz., Pennsylvania, New York, New England, and Newfoundland, have liberty to import salt from any part of Europe directly. The other eight, viz., Virginia, Maryland, East and West Jersey, North and South Carolina, Georgia and Nova Scotia, as well as all the West India Islands, are deprived of it.

At present those Colonies on whose behalf the petition is given, are supplied with salt from the Isle of Mays in Africa, Sal Tortuga, and Turks Island in America, also a little from England; but are deprived of the only salt that answers best for the principal use, viz., to preserve fish and other provisions, twelve months, or a longer time. What they have from Great Britain is made from salt water by fire, which is preferred for all domestic uses. The African or American salt is made from salt water by the sun; which is used for curing and preserving provisions. The first, made by fire, is found, by long experience, in warm climates, to be too weak; the provisions cured with it turn rusty, and in six or eight months become unfit for use. The second kind, by the quantity of alum, or some other vicious quality in it, is so corrosive, that in less than twelve months, the meat cured with it is entirely deprived of all the fat, and the lean hardened, or so much consumed, as to be of little service. The same ill qualities are found in these salts with regard to fish: wherefore the arguments used, that they ought to have English salt only, are as much as to say, they should be allowed to catch fish, or salt any provisions, but let their cattle and hogs die without reaping the advantage nature has given them.

In all countries where a benefit can arise by fish or provisions, salt must be cheap; and as its value where made is from ten to twenty shillings the ton, so the carriage of it to America is often more than the real value: It is in order to save part of the expense of carriage, this application is made; for although some gentlemen do not seem to know it, yet we have liberty, by the present laws in force, to carry any kind of European salt to America, the ship first coming to an English port, in order to make an entry.

We have also liberty to bring it from any salt island in Africa or America; but by the Act of 15 Car. II. Chap. 7, salt is supposed to be included under the word commodity; whereby it is, with all European goods, prevented from being carried to America, unless first landed in England: the consequence whereof is, that English ships, which (I shall suppose) are hired to sail from London to Lisbon with corn, and thence proceed to America, have not the liberty to carry salt in place of ballast, and therefore under a necessity to pay above £10 sterling at Lisbon for ballast (that is to say, for sand), which they carry to America, or else return to England in order to get a clearance for the salt, which would be more expense than its value.

Now, had they liberty to carry salt directly to America, they would not only save the money paid for the sand, but also gain by the freight of salt perhaps £60 or £80 more. Thus on an average every ship that goes now empty from these ports to America, might clear £70 and there are above a hundred sail to that voyage every year. This is an annual loss of £7,000 at least; and besides, as the ship loses no time in this case (salt being as soon taken in as sand), they could afford to sell the best salt as cheap in America as is now paid for the worst; for as a ship must make a long voyage on purpose to get, and make it in the salt islands, so the expense thereof is more than the value of the salt at Lisbon, St. Ibbes, and so forth.

The proponents of the petition made out a strong case. They went into the grading of the kinds of salt obtained from the West Indies, Africa and Europe and asserted that, inferior though some of them were, they nevertheless had been found to be "preferable to England salt for curing and preserving their fish":

To know the qualities of the different kinds of salt used in America may be an amusement to a speculative man; but seems entirely out of the question in this case; for whatever may be said on that head, long experience and the universal agreement of all from America, as well as former Acts of Parliament, show that the common white salt will not answer the uses it is chiefly wanted for there.

As to what is called Loundes's brine salt, that, and his many other projects, seemed to be formed on the same plan with Subtle's in *The Alchemist*, his scheme looking as if he only wanted the money, and left it to others to make the salt.

Salt can, without doubt, be made of any desired quality, but the price, the place of delivery, and the quantity to be had of so useful a commodity must also be regarded.

We can get salt at Sal Tortuga for the raking and putting it into our ships; but the expense of a voyage on purpose for it is greater than to buy it at a place from whence the freight may be all saved, and to have the best salt on the cheapest terms, is, no doubt the intention of this application, as it certainly was of the other Colonies that have obtained this privilege.

All the Virginians were asking, in effect, was the liberty to import from Europe what salt they wished!

As the moment of Independence neared, the stress grew greater. George Washington's Mt. Vernon overseer during the crucial years, his distant relative Lund Washington, addressed a letter to him in 1775:

The people are running mad about salt. You would hardly think it possible there could be such a scarcity. Five and six shillings per bushel. Conway's sloop came to Alexandria Monday last with a load.

A couple of months later the crisis was reached:

I have had 300 bushels more of salt put into fish barrels, which I intend to move into Muddy Hole barn, for if it should be destroyed by the enemy we shall not be able to get more. There is still fifty or sixty more bushels, perhaps a hundred in the house. I was unwilling to sell it, knowing we could not get more and our people must have fish. Therefore I told the people I had none.

Two more years of adversity went by. Lund wrote in 1778:

I was told a day or two past that Congress had ordered a quantity of shad to be cured on this river. I expect as everything sells high, shad will also. I should be fond of curing about 100 barrels of them, they finding salt. We have been unfortunate in our crops, therefore I could wish to make something by fish.

He proposed that he cure fish "for the Continent" and make "upwards of 200 pounds":

I have very little salt, of which we must make the most. I mean to make a brine and after cutting off the head and bellies, dipping them in the brine for but a short time, then hang them up and cure them by smoke, or dry them in the sun; for our people being so long accustomed to have fish whenever they wanted, would think it very bad to have none at all.

All ended well for that season. Lund wrote:

I have cured a sufficient quantity of fish for our people, together with about 160 or 170 barrels of shad for the Continent.

One of the most interesting diarists of Revolutionary days was young Nicholas Cresswell, an Englishman of 24 when he arrived in America for a three-years visit. He was in Leesburg, Virginia, in December 1776 when he recorded this occurrence:

A Dutch mob of about forty horsemen went through the town today on their way to Alexandria to search for salt. If they find any they will take it by force.... This article is exceedingly scarce; if none comes the people will revolt. They cannot possibly subsist without a considerable quantity of this article.

The raiders were pacified by an allotment of three pints of salt per man.

A vivid picture of what the lack of salt entailed was given by Cresswell in April 1777:

Saw a seine drawn for herrings and caught upwards of 40,000 with about 300 shad fish. The shads they use but the herrings are left upon the shore useless for want of salt. Such immense quantities of this fish is left upon the shore to rot, I am surprised it does not bring some epidemic disorder to the inhabitants by the nauseous stench arising from such a mass of putrefaction.

A fishery by-product of importance to early Virginians, lime, was of interest to Washington. It was extensively obtained by burning oyster shells.

Early Virginia masonry shows that such lime was mixed in mortar and it was usually of poor quality, perhaps because of crude facilities for burning. Today's shell lime is much in demand in agriculture and its price is higher than mined lime. George Washington found that for the purpose of building it left much to be desired. He wrote to Henry Knox from Mt. Vernon in 1785:

I use a great deal of lime every year, made of the oyster shells, which, before they are burnt, cost me twenty-five to thirty shillings per hundred bushels; but it is of mean quality, which makes me desirous of trying stone lime.

He was paying about seven cents a bushel for shells, which seems high for those days of abundant oysters and cheap labor. Until recently the Virginia market price was very little more.

Washington's probing, weighing mind slighted no phase of his fishery. About to fertilize crops with fish experimentally, he wrote to his overseer: "If you tried both fresh and salt fish as a manure the different aspects of them should be attended to." A few weeks later, after watching results, he wrote: "The corn that is manured with fish, though it does not appear to promise much at first, may nevertheless be fine.... It is not only possible but highly probable."

This opinion was abundantly confirmed years later when vast quantities of menhaden were converted into guano for crops by Atlantic coast factories, a practice changed only when livestock-nutrition studies showed that menhaden scrap was too valuable a protein source to be spread on land. The fish referred to by Washington were in all probability river-herring, or alewives, used as fertilizer at such times as they were caught in greater abundance than the food market could absorb.

The probable yield of his fish trade was always carefully calculated, even when the pressure of national affairs required his absence from home. From Philadelphia we find him writing to his manager about a fish merchant's offer: "Ten shillings per hundred for shad is very low. I am at this moment paying six shillings apiece for every shad I buy." He usually tried to get at least twelve shillings a hundred for his shad, which were salted prior to marketing, although there were instances when he let them go for as little as one pence apiece. The extraordinary price of six shillings for one shad cited by him in Philadelphia is hard to explain. It probably referred to a fresh one caught early in the season and prepared especially for his table. Though records of the average weight of shad in those days are lacking, seven pounds is a fair estimate, and it may have been greater. The weights now seldom exceed three or four pounds, because in the more recent years of intensive fishing, shad have been widely caught up as they returned from the ocean to spawn for the first time. Shad, along with other anadromous, or "up-running," fish are born near the head-waters of rivers, and seek the ocean for feeding and growth. Unlike salmon they do not perish after one spawning and the oftener they return, the larger they are. What conservationists call "escapement," or the freedom to get back to the ocean from the rivers, is considered vital to their survival in quantity.

All through the two-score years of fishing at Mount Vernon, Washington suffered, judging by his unceasing preoccupation with minor details, from the lack of a fishing foreman to whom he could entrust the operation with any confidence. Letters toward the close of his life bearing on this subject are still replete with reminders concerning trifles which would have been routine for any competent boss. The fish runs start about March; therefore, in January he finds it necessary to write; "It would be well to have the seines overhauled immediately, that is, if new ones are wanting, or the old ones requiring much repair, they may be set about without loss of time." He must even look beyond his own help for the skill necessary to put his nets in order. "I would have you immediately upon the receipt of this letter send for the man who usually does this work for me.... Let him choose his twine (if it is to be had in Alexandria) and set about them immediately."

Abundance of fish created a bottleneck:

In the height of the fishery they are not prepared to cure or otherwise dispose of them as fast as they could be caught; of course the seines slacken in their work, or the fish lie and spoil when that is the only time I can make anything by the seine, for small hauls will hardly pay the wear and tear of the seine and the hire of the hands.

However, then as now, fishing was a gamble:

Unless the weather grows warmer your fishing this season will, I fear, prove unproductive; for it has always been observed that in cold and windy weather the fish keep in deep water and are never caught in numbers, especially at shallow landings.

And in 1794, he states, with the rather weary voice of experience,

I am of opinion that selling the fish all to one man is best ... if Mr. Smith will give five shillings per thousand for herrings and twelve shillings a hundred for shad, and will oblige himself to take all you have to spare, you had better strike and enter into a written agreement with him.... I never choose to sell to wagoners; their horses have always been found troublesome, and themselves indeed not less so, being much addicted to the pulling down and burning the fences. If you do not sell to Smith the next best thing is to sell to the watermen.... I again repeat that when the schools of fish run you must draw night and day; and whether Smith is prepared to take them or not, they must be caught and charged to him; for it is then and then only I have a return for my expenses; and then it is the want of several purchasers is felt; for unless one person is extremely well prepared he cannot dispose of the fish as fast as they can be drawn at those times and if the seine or seines do no more than keep pace with his convenience my harvest is lost and of course my profit; for the herrings will not wait to be caught as they are wanted to be cured.

Thus did Washington become one of the first to encounter the besetting plague of American mass production: the problem of distribution.

That fishing was a vital prop in plantation economy is evidenced by a letter of April 24, 1796, to his manager:

As your prospect for gain is discouraging, it may, in a degree, be made up in a good fishing season for herrings; that for shad must, I presume, be almost, if not quite, over.

Salt herrings were a staple in the feeding of the "black people," and were issued to those at Mount Vernon at the rate of twenty a month per head. But he warned about waiting for the annually expected herring "glut" to occur before the slaves were provided for. If it should fail to materialize—as had been known—what then? Save a "sufficiency of fish" from the first runs, he wisely ordered.

In 1781 he suggested that salt fish be contracted for the troops, and possibly it was tried for a while, but the year following, army leaders voted to exclude fish from the rations.

Accounting records for 1774, presumably an average fishing year, show receipts of £170 for the catch at the Posey's ferry fishery, with £26 debited to operating cost. At the Johnson's ferry fishery £114 was taken in and £28 paid out. The catch here represented consisted of 9,862 shad and 1,591,500 river herring, but other large hauls were also made on the estate. Profits would seem to be adequate, although costs of nets and boats were not figured in. Fishing boats were usually small maneuverable craft that never had to put out very far from shore, and cost about £5 to build.

Occasionally Washington was approached by speculators offering to rent the season's privileges at one of his fisheries for a flat sum. About one such proposal in 1796 he expressed the opinion to his manager that "under all chances fishing yourself will be more profitable than hiring out the landing for £60." Nevertheless, the headaches had for years made the transference of fishing to someone for cash on the barrelhead a temptation. In February, 1770, he had entered into an agreement as to sales while retaining the responsibility of catching:

Mr. Robert Adams is obliged to take all I catch at Posey's landing provided the quantity does not exceed 500 barrels and will take more than this quantity if he can get casks to put them in. He is to take them as fast as they are catched, without giving any interruption to my people, and is to have the use of the fish house for his salt, fish, etc., taking care to have the house clear at least before the next fishing season; is to pay £10 for the use of the house and 3 shillings 4 pence, Maryland currency, per hundred for white fish.

But in 1787 he wrote: "A good rent would induce me to let the fishery that I have no trouble or perplexity about it." The *Diary* shows a good deal more interest during the early years in how the fish ran than it does later. In April, 1760, he writes:

Apprehending the herring were come, hauled the seine but catched only a few of them, though a good many of other sorts.... Hauled the seine again, catched two or three white fish, more herring than yesterday and a great number of cats.

August, 1768: Hauling the seine upon the bar of Cedar Point for sheepshead but catched none.

April, 1769: The white fish ran plentifully at my seine landing, having catched about 300 at one haul....

The term "white fish" is not now generally applied to any species caught in the Potomac, but a good guess is that, with Washington, it was an alternate for shad.

The Revolution was fought, but even before the surrender the minds of America's statesmen were actively considering peace terms. Both Richard Henry Lee and Thomas Jefferson suggested that the valuable fisheries off Newfoundland be freely open to American ships. This time it was not a question of the Northern Colony keeping the Southern Colony out as it had been 150 years before. Thomas Jefferson, writing in 1778, wanted the United Colonies to exclude England:

If they [Britain] really are coming to their senses at last, and it should be proposed to treat of peace, will not Newfoundland fisheries be worthy particular attention to exclude them and all others from them except our *très grand* and *chers amis* and allies? Their great value to whatever nation possesses them is as a nursery for seamen. In the present very prosperous situation of our affairs, I have thought it would be wise to endeavor to gain a regular and acknowledged access in every court in Europe but most the Southern. The countries bordering on the Mediterranean I think will merit our earliest attention. They will be the important markets for our great commodities of fish, wheat, tobacco, and rice.

Lee saw how fishing in Northern waters had started America on its way to being a maritime power. In a series of letters to George Mason and others he expresses his opinions forcibly:

Our news here is most excellent; both from Williamsburg and from Richmond it comes that our countrymen have given the enemy in the South a complete overthrow.... Heaven grant it may be so. I shall then with infinite pleasure congratulate my friend on the recovery of his property, and our common country on so great a step towards really putting a period to the war. I think that in this case we may insist on our full share of the fishery, and the free navigation of the Mississippi. These are things of very great and lasting importance to America, the yielding of which will not procure the Congress thanks either from the present age or posterity.

I rejoice greatly at the news from South Carolina. God grant it may be true. If this should force the enemy to reason and to peace, would you give up the navigation of the Mississippi and our domestic fishery on the Banks of Newfoundland? The former almost infinitely depreciating our back country and the latter totally destroying us as a maritime power. That is taking the name of independence without the means of supporting it.

I rejoice exceedingly at our successes both in the North and in the South. If we continue to do thus, it will not be in the power of the execrable junto to prevent us from having a safe and honorable peace next winter. In this idea I shall ever include the fisheries and the navigation of the Mississippi. These, Sir, are the strong legs on which North America can alone walk securely in independence.

If you do not get a wise and very firm friend to negotiate the fishery, it is my clear opinion that it will be lost, and upon this principle that it is the interest of every European power to weaken us and strengthen themselves.

I heartily wish you success in your negotiations and that when you secure one valuable point for us (the fishery) that you will not less exert yourself for another very important object, — the free navigation of the Mississippi, provided guilty Britain should remain in possession of the Floridas.

Fishing as a matter of states' rights resulted in the pioneering Potomac River Compact of 1785, when representatives of Maryland and Virginia met under George Washington's sponsorship at Mt. Vernon to deal with fishing and tolls. Maryland owned the river to the Virginia shore line, and agreed to allow Virginians to fish in it in return for free entry of Maryland ships through the Virginia capes. The compact, in force to this day, was the first step taken in behalf of interstate commerce. With its example to follow, other states eased the barriers to their commercial interests, with immeasurable benefit to the Union.

Commercial fishing in Virginia was, as the century closed, on the verge of the stability it had sorely lacked. Its reliance on Indians for knowledge and skill, as in the first of the 17th century, was as dead as its reliance on England for manufactures in the last of the 18th. Just around the corner were railroads and steamboats with their comparatively swift transportation. Teeming cities needed to be fed, and after nearly two centuries of education in the ways of the Chesapeake Bay and its marine life, Virginia fishermen knew how to keep the markets stocked. In 1794 a French visitor, Moreau de Saint Méry, wrote:

Fish is the commodity that sells for a ridiculously low price in Norfolk. One can purchase weakfish weighing more than twenty pounds for 4 or 5 francs and sometimes one that weighs three times more for a gourde, 5 francs, 10 sous. Drum is also very cheap. Sturgeon, weighing up to 60 pounds, can be bought for 6 French sous a pound, about the same price paid for little codfish that are brought in alive and are delicious to eat. Shad is also plentiful there. In addition, one can get perch, porpoise, eels, leatherjackets, summer flounder, turbot, mullet, trout, blackfish, herring, sole, garfish, etc. In short, fish is so abundant in Norfolk that sometimes the police find it necessary to throw back into the water those that are not bought.

Herring fishing began to be abandoned by the planters, many of whom were up to their necks in a variety of enterprises, in favor of business men intending to specialize. Letters from a Virginia speculator, John F. Mercer, to Richard Sprigg, sketch the situation:

April 19, 1779. To cure fish properly requires two days in the brine before packing and they can only lie packed with safety in dry weather. These circumstances joined with the heading and drawing almost all the fish (a very tedious operation) will show that no time was lost—only 9 days elapsed from his arrival here to his completing his load of 15,000 herrings, a time beyond which many wagons have waited on these shores for 4,000 uncured fish and many have been obliged to return without one, after coming 40 and 50 miles and offering 2 and 5 dollars a thousand. Several indeed from my own shore and six who want 36,000 herring will, I believe, quit this night without a fish, after waiting all this storm on the shore five days.

Mr. Clarke has had his fish completed two days.... He has been delayed by the almost continual storm that has prevailed since his arrival and which has ruined us fishermen.

My fishery has been miserably conducted from the beginning as might be expected from my entire ignorance and the penury of my partner who was poorer than myself.... Still I have expectations that it may turn out an immense thing from the trial we have made. The shores being opposite to Maryland Point, the reach above and below with the mouths of the two creeks on this side form a sweep, both tides upon them, that must collect for fish; and they are kept in by a kind of pound on the Virginia shore's trend. There apparent advantages accord with the experiment for, with a desperate patched-up seine that always breaks with a good haul, we have contrived to land 20,000 a day, every day we can haul. We are nearer to the Fredericksburg and Falmouth Virginia markets than any shore that is or can be opened on the river by 10 miles notwithstanding every discouragement and particularly the activity and lies practiced against us by the Little Creek fisheries on each side, who must fail with our success.

April 10, 1795. Herrings they tell me are 10 shillings per thousand at all the shores. If I had your lease I could make a fortune. I have a great mind to send Pollard and George up for your small boat and seine.... If Peyton comes down with his seine to haul at my shore, I will seine salted herrings enough for us both.

That acidulous but always colorful roving reporter from the midwest, Anne Royall, offers the best picture, for accuracy and detail, of hauling a seine ever presented by anyone not a technician. Though written almost 50 years after the Revolution, it describes the kind of fishing on which Virginians had principally depended since Christopher Newport began the Colonial era and George Washington ended it:

The market of Alexandria is abundant and cheap; though much inferior to any in any part of the western country, except beef and fish, which are by far superior to that of the western markets.... Their exquisite fish, oysters, crabs, and foreign fruits upon the whole bring them upon a value with us.

Their fish differ from ours, even some species. Their catfish is the only sort in which we excel; they have none that answer to our blue cat, either in size or flavor, and nothing like our mud-cat. Their catfish is from ten to fifteen inches in length, with a wide mouth, like the mud-cat of the Western waters; but their cat differ from both ours in substance and color; they are soft, pied black and white. They are principally used to make soup, which is much esteemed by the inhabitants. All their fish are small compared with ours. Besides the catfish which they take in the latter part of the winter, they have the rock, winter shad, mackerel, and perch, shad and herring. The winter shad is very fine indeed. They are like our perch, but infinitely smaller. These fish are sold very low; a large string, enough for a dozen persons, may be purchased for a few cents. No fish, however, that I have tasted, equal our trout.

The Potomac at Alexandria, is rather over a mile in width; it is celebrated for its beauty. It is certainly a great blessing to this country in supplying its inhabitants with food in the article of fish.

Fish is abundant (at Washington), and cheap at all seasons, shad is three dollars per hundred; herrings, one dollar per thousand.

Great quantities of herring and shad are taken in these waters during the fishing season, which commences in March, and lasts about ten weeks. As many as 160,000 are said to be caught at one haul. When the season commences no time is to be lost, not even Sunday. Although I am not one of those that make no scruple of breaking the Sabbath, yet, Sunday, as it was, I was anxious to see a process which I had never witnessed—I mean that of taking fish with a seine—there being no such thing in the Western country. It is very natural for one to form an opinion of some sort respecting things they have never seen, but the idea I had formed of the method of fishing with a seine was far from a correct one. In the first place, about fifteen or twenty men, and very often an hundred, repair to the place where the fish are to be taken, with a seine and a skiff. This skiff, however, must be large enough to contain the net and three men—two to row, and one to let out the net. These nets, or seines, are of different sizes, say from two to three hundred fathom in length, and from three to four fathom wide. On one edge are fastened pieces of cork-wood as large as a man's fist, about two feet asunder, and on the opposite edge are fastened pieces of lead, about the same distance—the lead is intended to keep the lower end of the seine close to the bottom of the river. The width of the seine is adapted to the depth of the river, so that the corks just appear on its surface, otherwise the lead would draw the top of the seine under water, and the fish would escape over the top. All this being understood and the seine and rowers in the boat, they give one end of the seine to a party of men on the shore, who are to hold it fast. Those in the boat then row off from the shore, letting out the seine as they go; they advance in a straight line towards the opposite shore, until they gain the middle of the river, when they proceed down the stream, until the net is all out of the boat except just sufficient to reach the shore from whence they set out, to which they immediately proceed. Here an equal number of men take hold of the net with those at the other end, and both parties commence drawing it towards the shore. As they draw, they advance towards each other, until they finally meet, and now comes the most pleasing part of the business. It is amusing enough to see what a spattering the fish make when they find themselves completely foiled: they raise the water in a perfect shower, and wet every one that stands within their reach. I ought to have mentioned, that when the fish begin to draw near the shore, one or two men step into the water, on each side of the net, and hold it close to the bottom of the channel, otherwise the fish would escape underneath. All this being accomplished, the fishermen proceed to take out the fish in greater or less numbers, as they are more or less fortunate. These fishermen make a wretched appearance, they certainly bring up the rear of the human race. They were scarcely covered with clothes, were mostly drunk, and had the looks of the veriest sots on earth.

A Virginian born in 1792, Col. T. J. Randolph of Edgehill near Charlottesville, was asked to search his earliest memories in order to record 18th century fishing conditions. He wrote a letter in 1875 to the newly-constituted Virginia fish commissioners describing an era well-nigh incredible to today's Tidewater fishermen:

Shad were abundant in the Rivanna at my earliest recollection, say prior to 1800. They penetrated into the mountains to breed. I have heard the old people, when I was young, speak of their descending the rivers in continuous streams in the fall, as large as a man's hand. The old ones so weak, that if they were forced by the current against a rock they got off with difficulty. Six miles north of Charlottesville three hundred were caught in one night with a bush seine. A negro told me he had caught seventeen in a trap at one time. I recollect the negroes bringing them to my mother continually. An entry of land near Charlottesville about 1735 crossed the Rivanna for two or three acres as a fishing shore. The dams absolutely stopped them, but they had greatly declined before their erection. In 1810 every sluice in the falls at Richmond was plied day and night by float seines. I never heard of rockfish above the falls, and supposed they were confined to Tidewater.... Rockfish were hunted on the Eastern Shore on horseback with spears. The large fish coming to feed on the creek shores, overflowed by the tide, showed themselves in the shallow water by a ripple before them. They were ridden on behind and forced into water too shallow for them to swim well, and were speared. I inferred from this fact that they confined themselves to the Tidewater. When young, I have heard the old people speak of an abundance of other fish. The supposition was that the clearing of the country, and consequent muddying of the streams, had destroyed them.

By sluicing the dams, and prohibiting fishing in sluices, or trapping, or anything that should bar their progress, I do not see why the shad should not return.

The shad have never returned to the up-country. But they still visit the vast inland waters below the Fall line, sometimes so abundantly that the price declines, as it did so recently as 1956, to where the fishermen can scarcely make a profit. Other fish referred to by the first Virginians continue to return, and will do so as long as our outreaching civilization does not deprive them of the natural conditions they need for survival.

The years closely following the Revolution brought profound readjustment in American commerce. Observations on whaling, a minor but vital home industry, filled many pages of a 1788 communication of Thomas Jefferson to John Jay, one of his confreres in the shaping of national policy. After sketching the uses of whale oil, its economic position and its history, he took up the particular problem facing the people of Nantucket, perhaps the foremost whalers in America. As long as they had been subjects of the British Empire they had been able to sell their oil duty-free in England. Now as aliens they must pay the same tariff charged other foreign traders. This meant the difference between a profitable and unprofitable enterprise. A few Nantucket seamen had even transferred to Nova Scotia in order to become British citizens again and thus receive exemption from whale-oil import duty. This trend alarmed the French in particular, who could visualize thousands of the United States' best sailors going over to their enemies the English. The remedy was suggested: make France the most attractive market for U.S.

whale oil. At the same time, English whaling had been government subsidized and could undercut competition.

The international chess game went briskly on, to the concern of Jefferson and the well-wishers of the infant Union. Before the Revolution England had fewer than 100 vessels whaling, while America had more than 300. But by 1788 England had 314 and America 80. Such was the result of the conflict, aided by the bounty paid by Britain to its own whalers. Jefferson hoped that the United States producers could develop a market in France, in part, by bartering oil for the essential work clothes which hitherto had been bought for cash in England. But he warned that without some kind of subsidy American whalers could neither compete with foreign countries nor make a living commensurate with other pursuits. The growing nation's sea-faring men would decrease to the point where the country's sea power would be in question.

As Secretary of State in 1791, Jefferson reported to Congress on the two principal American fisheries of the day, both oceanic. "The cod and whale fisheries," he began, "carried on by different persons, from different ports, in different vessels, in different seas, and seeking different markets, agree in one circumstance, as being as unprofitable to the adventurer as important to the public." Once prosperous, he said, they were now in embarrassing decline.

He traced the history of the cod fisheries back to 1517, in which year as many as 50 European ships were reported fishing off the Newfoundland banks at one time. In 1577 there were 150 French vessels, 100 Spanish and 50 Portuguese. The British limped far behind with 15. The French gradually took over as they claimed more and more territory in the region. Other nations dropped out, except England, whose cod fleet at the beginning of the seventeenth century had increased to about 150 vessels. These in due course were largely supplanted by the New England colonists. When France lost Newfoundland to England in 1713 the English and Colonial fisheries spurted ahead. By 1755 their fleets and catches equaled those of the French, and in 1768 passed them. Jefferson's statistics present an impressive picture of the fishing activity of that time and place, especially when compared with the unorganized Chesapeake fisheries just then coming of age.

In 1791 he said there were 259 French vessels totaling 24,422 tons and employing 9,722 seamen. Their catch: 20 million pounds that year. There were 665 American vessels with 25,650 tonnage, 4,405 seamen and a catch of around 40 million pounds. England's ships, tonnage and men were not given. However, her estimated catch nearly equaled that of France and America combined. Thus the Northern fishing grounds in their palmy days accounted for well over 100 million pounds of cod a year.

It is worth remarking that the size of today's New England cod fishery is not radically different from the pre-Revolutionary one described by Jefferson. Boats, men and catch remain about the same on the average.

Turning to the whaling industry, Jefferson noted that Americans did not enter it until 1715, although he credited the Biscayans and Basques of Southern Europe with prosecuting it in the 15th century and leading the way to the fishing grounds off Newfoundland. Whales were sought in both the North and South Atlantic. The figures for the American Colonies in 1771 as given by Jefferson were 304 vessels engaged, totaling 27,800 tons, navigated by 4,059 men.

They were in for a difficult time in 1791. The Revolution halted their activities and deprived them of their markets. Re-establishing this fishery was a prime concern of Jefferson.

It is significant that in his painstaking consideration of the nation's fisheries he, a Virginian, apparently found no cause to deal with those of his own Chesapeake bay. They were one day nevertheless to outstrip many times over both the volume and value of American cod and whale fisheries together.

The evidence is that Jefferson was more interested in fish at Monticello than anywhere else. But there the interest was personal, not national. In his so-called *Farm Book*, or plantation record, he often mentions fish. A note on slave labor reads: "A barrel of fish costing $7. goes as far with the laborers as 200 ponds of pork costing $14." This was in all probability Virginia salt-herring, which had finally reached the status of a staple during the latter half of the 18th century. An 1806 memorandum to his overseer runs: "Fish is always to be got in Richmond ... and to be dealt out to the hirelings, laborers, workmen, and house servants of all sorts as has been usual." In 1812 a bill for fish, which he terms "indeed very high and discouraging, but the necessity of it is still stronger," lists the species no doubt in chief demand: "Twelve barrels herrings, $75. and one barrel of shads, $6.50." These were salted and shipped in from Tidewater fisheries like George Washington's at Mt. Vernon.

For fresh fish Jefferson and his neighbors could look to their adjacent rivers. In fact, so greatly did they rely on them that it was with feelings akin to consternation that he wrote his friend William D. Meriwether in 1809 that a neighbor, Mr. Ashlin, proposed to erect a dam which was sure to inconvenience the watermen of the vicinity. Furthermore, "to this then add the removal of our resort for fresh fish ... and the deprivation of all the intermediate inhabitants who now catch them at their door." He was not on too firm ground in objecting, however. He had a dam of his own across the Rivanna river which had been there since 1757.

He decided to build a fish pond in his garden. As he described it in 1808 it was little larger than an aquarium, 40 cubic yards contents, probably for water lilies and goldfish. It was the first of several fish ponds, constructed, no doubt, with both beauty and utility in mind. A note in his *Weather Memorandum Book* under date April 1812 tells us: "The two fish ponds on the Colle branch were 40 days work to grub, clean and make the dams."

A series of letters in 1812 to friends who he thought might supply him with live fish, particularly carp, for stocking, all run very much on the order of this one to Captain Mathew Wills:

I return you many thanks for the fish you have been so kind as to send me, and still more for your aid in procuring the carp, and you will further oblige me by presenting my thanks to Capt. Holman & Mr. Ashlin. I have found too late, on enquiry that the cask sent was an old and foul one, and I have no doubt that must have been the cause of the death of the fish. The carp, altho it cannot live the shortest time out of water, yet is understood to bear transportation in water the best of any fish whatever. The obtaining breeders for my pond being too interesting to be abandoned, I have had a proper smack made, such as is regularly used for transporting fish, to be towed after the boat, and have dispatched the bearer with it without delay, as the season is passing away. I have therefor again to solicit your patronage, as well as Captain Holman's in obtaining a supply of carp. I think a dozen would be enough and would therefore wish him to come away as soon as he can get that number.

From that time on his ponds came in for periodic mention, as when one was broken up by flood waters in 1814. But despite setbacks he kept faith in them as good food-producing adjuncts of a farm, thus anticipating the U.S. Department of Agriculture's modern food-fish pond-development program by more than a century.

As is likely to be the case with experimenters, Jefferson's efforts at fish propagation do not appear to have been overwhelmingly successful. At any rate, there is much more frequent reference in his records to putting fish in his ponds than taking them out. So far as he was concerned, it may be said that results were less important than example. Like all great leaders he was an originator and investigator, confining himself to the basic things that insure man's sustenance and contribute to his happiness, not the least of which is fishing.

BIBLIOGRAPHY

Archer, Gabriel. *A Relation of the Discovery of Our River From James Forte into the Maine, Made by Captain Christopher Newport.* Worcester, 1860.

Beverley, Robert. *The History and Present State of Virginia.* London, 1705.

Brown, Alexander. *The Genesis of the United States.* Boston, 1890. 2 vols.

Burnaby, Andrew. *Travels Through the Middle Settlements in North America in the Years 1759-1760.* London, 1798.

Byrd, William. *Natural History of Virginia.* Ed. and tr. by R. C. Beatty and W. J. Mulloy. Richmond, 1940.

Chastellux, François J. *Travels in North America in the Years 1780, 1781, and 1782.* London, 1787.

Cresswell, Nicholas. *The Journal, 1774-77.* Ed. by Lincoln McVeagh. New York, 1924.

De Vries, David P. *Voyages From Holland to America, 1632-1644.* New York, 1857.

Durand, —. *A Huguenot exile in Virginia.* Ed. by Gilbert Chinard. New York, 1934.

Fithian, Philip V. *Journal and Letters, 1773-1774.* Ed. by Hunter D. Farish. Williamsburg, 1943.

Force, Peter. *Tracts and Other Papers.* Washington, 1836-46. 4 vols.

Glover, Thomas. *An Account of Virginia.* London, 1676.

Hamilton, Stanislaus M., ed. *Letters to Washington and Accompanying Papers.* Boston, 1898-1901. 5 vols.

Hamor, Ralph. *Notes of Virginian affaires of the Government of Sir Thomas Gates and of Sir Thomas Dale till 1614.* Glasgow, 1906.

— — *A True Discourse of the Present State of Virginia.* London, 1614.

Hariot, Thomas. *Narrative of the First English Plantation of Virginia.* London, 1893.

Hart, Albert B. *American History Told by Contemporaries.* New York, 1908. 4 vols.

Hening, William W. *The Statutes at Large of Virginia.* 1809-1823. 13 vols.

Jefferson, Thomas. *The Complete Jefferson.* Ed. by Saul K. Padover. New York, 1943.

— — *Thomas Jefferson's Farm Book.* Ed. by Edwin M. Betts. Princeton. 1953.

— — *Thomas Jefferson's Garden Book, 1766-1824.* Ed. by Edwin M. Betts. Philadelphia, 1944.

Lee, Richard Henry. *Letters of Richard Henry Lee.* Ed. by James C. Ballagh. New York, 1914. 2 vols.

Middleton, Arthur P. *Tobacco Coast.* Ed. by George C. Mason. Newport News, 1953.

Neill, Edward. *Virginia Vetusta.* Albany, 1885.

Newport, Christopher. *A Description of the Now-discovered River and Country of Virginia, 1607.* Worcester, 1860.

Pearson, John C. *The Fish and Fisheries of Colonial Virginia.* In William and Mary College Quarterly, 1942-3. Williamsburg.

Purchas, Samuel. *His Pilgrimes.* Glasgow, 1906. 20 vols.

Royall, Anne. *Sketches of History, Life and Manners in the United States.* New Haven, 1826.

Smith, John. *Travels and Works of Captain John Smith.* Ed. by Edward Arber. Edinburgh, 1910. 2 vols.

Strachey, William. *The Historie of Travaile Into Virginia Britannia.* London, 1849.

Swem, E. G. *Virginia Historical Index.* Roanoke, 1934-6. 2 vols.

Virginia. *Calendar of Virginia State Papers.* Richmond, 1875-1893. 11 vols.

Virginia Fish Commissioners. *Annual Report for the Year 1875.* Richmond, 1875.

Virginia Company. *The Records*. Ed. by S. M. Kingsbury. Washington, 1906-1935. 4 vols.

Washington, George. *The Writings of George Washington*. Ed. by J. C. Fitzpatrick. Washington. 39 vols.

Whitelaw, Ralph T. *Virginia's Eastern Shore*. Ed. by George C. Mason. Richmond, 1951. 2 vols.

Manuscripts

Mercer Papers, Virginia Historical Society, Richmond.

Washington, Lund. *Letters*. Unpublished, at Mt. Vernon.

www.ingramcontent.com/pod-product-compliance
Lightning Source LLC
Chambersburg PA
CBHW030444220526
45464CB00006B/2405